Alvah Hunt Doty

A Manual of Instruction in the Principles of Prompt Aid to the Injured

Including a Chapter on Hygiene and the Drill Regulations for the...

Alvah Hunt Doty

A Manual of Instruction in the Principles of Prompt Aid to the Injured
Including a Chapter on Hygiene and the Drill Regulations for the...

ISBN/EAN: 9783337021764

Printed in Europe, USA, Canada, Australia, Japan

Cover: Foto ©berggeist007 / pixelio.de

More available books at **www.hansebooks.com**

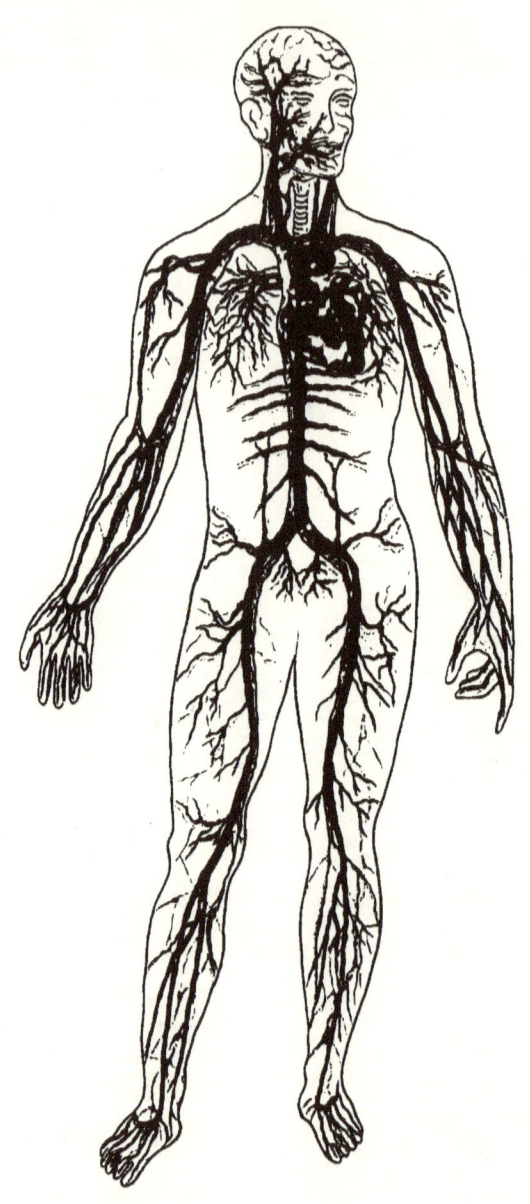

GENERAL PLAN OF THE CIRCULATION.

IN THE PRINCIPLES OF

PROMPT AID TO THE INJURED

INCLUDING A CHAPTER ON HYGIENE AND THE
DRILL REGULATIONS FOR THE HOSPITAL CORPS, U. S. A.

DESIGNED FOR MILITARY AND CIVIL USE

BY

ALVAH H. DOTY, M.D.

HEALTH OFFICER OF THE PORT OF NEW YORK
LATE MAJOR AND SURGEON, NINTH REGIMENT, N. G. S. N. Y.
LATE ATTENDING SURGEON TO BELLEVUE HOSPITAL DISPENSARY, NEW YORK

THIRD EDITION, REVISED AND ENLARGED

NEW YORK
D. APPLETON AND COMPANY
LONDON: 25 BEDFORD STREET
1898

TO

JOSEPH D. BRYANT, M. D.,

IN ACKNOWLEDGMENT OF HIS HIGH POSITION

AS A TEACHER OF ANATOMY AND SURGERY,

AND AS A LUCID WRITER

AND EXPONENT OF THEIR PRACTICAL APPLICATION,

THIS MANUAL IS DEDICATED

BY THE AUTHOR.

PREFACE TO SECOND EDITION.

IN preparing the second edition of "Prompt Aid to the Injured" I have endeavored to make such changes as will cause the subject-matter to conform to the present knowledge of the different topics included in this work.

A chapter on Hygiene has been introduced, with the hope that it will give to the reader a general idea of the methods by which the body may be kept in a healthy state. A chapter on Nursing was contemplated, but the space available for this purpose was insufficient to properly present the subject; it was therefore omitted.

Through the courtesy of the Surgeon-General of the U. S. Army I am permitted to add the recently adopted drill regulations for the Ambulance Corps of this service.

ALVAH H. DOTY, M. D.

59 WEST THIRTY-FIFTH STREET, NEW YORK,
    March 11, 1894.

PREFACE.

The object of this Manual is to instruct those who are desirous of knowing what course to pursue in emergencies, in order that sick or injured may be temporarily relieved. Special effort has been made to so arrange the matter and introduce such points as will be of use to the ambulance corps connected with the different military organizations. It will be appreciated that it is a difficult task to give to non-medical persons information which will properly instruct them to cope with emergencies without encouraging them to usurp the functions of the physician or surgeon.

In order that the subject may be well understood, it is essential to know something of the construction of the human body and the functions of the different organs; for this reason, considerable space has been devoted to anatomy and physiology. The author has endeavored to explain each topic in a plain and simple manner, and when medical terms are used their lay synonyms are also given. Numerous illustrations have been inserted to aid in making the work the more intelligible—many of which have been taken from the works of Esmarch, Flint, Tracy, and others, due credit, however, being given in each instance.

The author is particularly indebted to Dr. Glover C. Arnold, of this city, for proof-reading, and also many valuable suggestions. To Major Charles Smart, Surgeon, U. S. A., the author wishes to express his thanks for the permission given to introduce in this work the Manual of Transportation now used by the U. S. Army.

ALVAH H. DOTY, M. D.

59 WEST THIRTY-SIXTH STREET, NEW YORK,
    March 4, 1889.

CONTENTS.

CHAPTER I.

BONE 1
    Bone: its composition and function—Periosteum and endosteum—Classification of bone—The skeleton—Spine—Bony landmarks—Skull—Hyoid bone—Thorax—Sternum and ribs—Scapula—Clavicle—Bones of upper extremity—Pelvis—Bones of lower extremity.

CHAPTER II.

JOINTS—CARTILAGE—LIGAMENTS—SYNOVIAL MEMBRANE—MUSCLES . 23
    Joints: their classification and movement—Cartilage—Ligaments—Synovial Membrane—Muscles—Tendons and aponeuroses—Definition of organs—Glands—Mucous and serous membrane—Excretion and secretion.

CHAPTER III.

THE BLOOD AND CIRCULATORY ORGANS 30
    Blood: its composition and function—Heart—Pericardium—Circulation of the blood—Endocardium—Blood-vessels.

CHAPTER IV.

RESPIRATION 39
    Definition of respiration—Larynx—Trachea—Bronchial tubes—Lungs.

CHAPTER V.

ALIMENTATION AND DIGESTION 45
    Alimentary tract—Mastication—Teeth—Salivary glands—Pharynx—Œsophagus—Stomach—Small and large intestine—Liver—Pancreas.

CHAPTER VI.

KIDNEYS—BLADDER—SKIN—SPLEEN 55

CHAPTER VII.

NERVOUS SYSTEM 59
 Cerebro-spinal system — Brain — Cerebrum — Cerebellum — Pons Varolii—Medulla oblongata—Spinal cord—Cranial and spinal nerves—Sympathetic system.

CHAPTER VIII.

BANDAGES AND DRESSINGS 68
 Bandages: their classification — Material for construction — Method of application—Roller, Esmarch, or handkerchief bandages—Slings—Knots—Compresses—Tampons—Poultices—Moist and dry heat.

CHAPTER IX.

ANTISEPTICS—DISINFECTANTS—DEODORANTS 91
 Manner of using—Rules of Health Department, New York city—Sterilization of milk for infants.

CHAPTER X.

CONTUSIONS AND WOUNDS 109
 Classification and treatment.

CHAPTER XI.

HÆMORRHAGE 120
 Hæmorrhage—Arterial—Venous—Capillary—Means of arresting hæmorrhage, and how applied.

CHAPTER XII.

FRACTURES, DISLOCATIONS, SPRAINS 136
 Fractures: their classification—Signs of fracture—Method of repair by Nature—Treatment of special fracture—Dislocations—Sprains.

CHAPTER XIII.

BURNS, SCALDS, AND FROST-BITE 156
 Burns: their classification—Treatment of each degree—Scalds and frost-bite, and their treatment.

CHAPTER XIV.

UNCONSCIOUSNESS, SHOCK OR COLLAPSE, AND SYNCOPE OR FAINTING 161
 Definition of, and treatment.

CHAPTER XV.

CONCUSSION AND COMPRESSION OF THE BRAIN—APOPLEXY, OR STROKE OF PARALYSIS — INTOXICATION — EPILEPSY, HYSTERIA, AND HEATSTROKE OR SUNSTROKE 169
Definition of, and treatment.

CHAPTER XVI.

ASPHYXIA AND DROWNING 178
Asphyxia, or suffocation—Treatment—Prevention—Drowning—Artificial respiration—Different methods—Sylvester's, Howard's, and Hall's.

CHAPTER XVII.

POISONS AND POISONING 186
Varieties of poisons—Narcotics, irritants, and corrosives—Antidotes and treatment.

CHAPTER XVIII.

CONVULSION OF CHILDREN—TETANUS—FOREIGN BODIES IN EYE, EAR, NOSE, LARYNX, AND PHARYNX—BED-SORES—FOOT-SORENESS AND CHAFING 199
Convulsions of children and treatment—Tetanus, or locked-jaw, and treatment—Foreign bodies in eye, ear, nose, larynx, and pharynx—Method of removal—Bed-sores, prevention and treatment—Foot-soreness and treatment—Chafing and treatment.

CHAPTER XIX.

HYGIENE 207
Baths—Clothing—Food—Water—Air—Exercise.

CHAPTER XX.

TRANSPORTATION OF THE WOUNDED 222
Litters or stretchers—Essential of litters—Varieties—Halstead's—Marsh's—Extemporized litters—Drill Regulations for the Hospital Corps, U. S. Army.

ILLUSTRATIONS.

FIGURE		PAGE
28. Abdominal contents, position of........................	*Flint.*	50
24. Air-cells and terminal bronchial tubes.................	*Flint.*	43
66. Arteries, diagram showing position of important.		
	Pye, modified. Face	125
70. Artery, brachial, digital compression of.............	*Esmarch.*	131
69. Artery, brachial, line showing course of...............	*Tracy.*	130
67. Artery, common carotid, digital compression of.		
	Esmarch, modified.	128
73. Artery, femoral, compression by tourniquet.........	*Esmarch.*	134
72. Artery, femoral, digital compression of.............	*Esmarch.*	133
71. Artery, femoral, line showing course of...............	*Tracy.*	132
68. Artery, subclavian, digital compression of.	*Esmarch, modified.*	129
84. Artificial respiration: Hall's method; first position..	*Original.*	185
85. Artificial respiration: Hall's method; second position.	*Original.*	185
82. Artificial respiration: Howard's method; first part..	*Original.*	183
83. Artificial respiration: Howard's method: second part.	*Original.*	184
80. Artificial respiration: Sylvester's method; first movement.	*Original.*	182
81. Artificial respiration: Sylvester's method; second movement.....................................	*Original.*	182
58. Bandages, cravat, for hand................	*Esmarch, modified.*	84
59. Bandages, cravat, for knee.........................	*Esmarch.*	84
47. Bandages, diagram of triangular.....................	*Original.*	77
44. Bandages, four-tailed, for head......................	*Esmarch.*	75
43. Bandages, four-tailed, for jaw.......................	*Original.*	75
39. Bandages, method of rolling.		
	Reference Hand-book of Medical Sciences.	69
42. Bandages, knotted.................................	*Esmarch.*	74
45, 46. Bandages, large square handkerchief..........	*Esmarch.*	76, 77
41. Bandages, spica of shoulder.		
	Reference Hand-book of Medical Sciences.	73

xiv ILLUSTRATIONS.

FIGURE		PAGE
40. Bandages, spiral reverse.		
	Reference Hand-book of Medical Sciences.	71
51. Bandages, triangular, for chest, etc........	*Esmarch, modified.*	80
57. Bandages, triangular, for foot.............	*Esmarch, modified.*	83
48, 49. Bandages, triangular, for head.........	*Esmarch, modified.*	78
56. Bandages, triangular, for hip.............	*Esmarch, modified.*	83
50. Bandages, triangular, for shoulder, hand, etc.	*Esmarch, modified.*	79
52. Bandages, triangular, for shoulder, head, etc.	*Esmarch, modified.*	80
1. Bone, cancellous and compact tissue, arrangement of...	*Tracy.*	2
33. Brain inclosed in membranes......................	*Didama.*	59
35. Brain, under surface of.............................	*Tracy.*	61
36. Brain, upper surface of.............................	*Didama.*	62
37. Brain and spinal cord.............................	*Tracy.*	65
9. Clavicle (collar-bone).............................	*Original.*	14
75. Clavicle, dressing for fracture of...............	*Pye, modified.*	145
18. Corpuscles of human blood, red and white............	*Flint.*	31
87. Extemporized stretcher............................	*Original.*	225
76. Femur, fracture of; gun used for temporary splint.		
	Esmarch, modified.	149
13. Femur (thigh-bone).............................	*Original.*	19
15. Foot, bones of.................................	*Original.*	21
11. Forearm (radius and ulna), bones of...............	*Original.*	16
21. Frog's foot, web of, magnified......................	*Flint.*	37
12. Hand, bones of.................................	*Original.*	17
19. Heart..	*Didama.*	32
20. Heart and ribs, relation of........................	*Flint.*	33
10. Humerus (arm-bone)............................	*Original.*	15
25. Jaws and teeth...................................	*Didama.*	46
31. Kidney, vertical section of.......................	*Flint.*	55
77. Knot, Gerdy's extension........................		150
63. Knot, granny...................................	*Bryant.*	85
62. Knot, granny, handkerchief.....................	*Esmarch.*	85
60. Knot, reef.....................................	*Bryant.*	85
61. Knot, reef handkerchief.........................	*Esmarch.*	85
64. Knot, surgeon's................................	*Bryant.*	86
14. Leg, bones of..................................	*Original.*	20
79. Leg, fracture of; pillow for temporary splint........	*Original.*	152
78. Leg, fracture of; umbrella for temporary splint.....	*Original.*	151
29. Liver...	*Tracy.*	51
86. Marsh stretcher................................	*Original.*	224
17. Muscular system................................	*Tracy.*	26
34. Nerves, cerebro-spinal system...................	*Tracy.*	60
38. Nerves, sympathetic system......................	*Tracy.*	66

ILLUSTRATIONS.

FIGURE		PAGE
30. Pancreas (sweet-breads)...............................	Tracy.	53
16. Patella (knee-cap).................................	Original.	22
7. Rib, a...	Original.	12
26. Salivary glands..................................	Tracy.	47
8. Scapula (shoulder-blade)........................	Original.	13
2. Skeleton...	Tracy.	5
32. Skin, vertical section of.........................	Didama.	56
4. Skull..	Original.	9
74. Sling, sleeve of coat used as a...............	Esmarch, modified.	140
54. Slings, triangular, for arm...................	Esmarch, modified.	81
55. Slings, triangular, for arm...................	Esmarch, modified.	82
53. Slings, triangular, for arm...................	Esmarch, modified.	81
3. Spine, vertical section of......................	Tracy.	6
27. Stomach..	Didama.	48
65. Suspender, Esmarch's..........................	Original.	124
5. Thorax (chest), anterior view..................	Sappey.	10
6. Thorax (chest), posterior view.................	Sappey.	11
23. Thorax (chest), cavity of, showing position of heart and lungs....................................	Didama.	41
22. Trachea and bronchial tubes....................	Sappey.	40

DRILL REGULATIONS FOR THE HOSPITAL CORPS, U. S. ARMY.

PAR. 18. Right oblique...	230
21. Detachment right..	231
24. Fours right..	233
29. Right forward, fours right.............................	235
34. On right into line.......................................	237
35. Right front into line....................................	238
38. Right by twos...	239
42. Form fours, left oblique................................	240
57. The vertical position...................................	243
58. Order litter...	243
59. Shoulder litter...	244
62. Carry litter...	245
66 and 71. At litter posts with open litter...............	247
67. Sling secured...	247
69. Strap litter...	248
88. Litters right..	252
103. Posts at patient..	255
108. The patient lifted.....................................	257
121. Passing an obstacle....................................	260
125. By four, carry litter..................................	261

	PAGE
PAR. 128. Carrying a loaded litter upstairs	263
129. Carrying a loaded litter downstairs	264
131. Carrying by three bearers	265
144. Lifting the patient erect	269
147. Patient across back	270
148. Patient across shoulder	271
150. Two-handed seat	272
153. The rifle seat	274
157. The travois	276
167. Loading ambulance	281
197. Scheme for packing medical officer's orderly pouch	294
198. Scheme for packing hospital corps pouch	295
199. Hospital corps bugle call	295

PROMPT AID TO THE INJURED.

CHAPTER I.

BONE.

BONE is composed of animal and earthy matter ; the former consists of gelatin, fat, and blood-vessels, and the latter, known as "bone-earth," is principally composed of phosphate of lime. Bone derives its tenacity and elasticity from the animal portion, and the "bone-earth" gives to it the necessary hardness.*

When soup is made by boiling a bone, there is simply a liberation of the gelatin or animal matter, which principle may be extracted, by the same process, even from bone two or three thousand years old. In early life the animal part forms about one third and the earthy matter two thirds of bone ; later in life, however, the proportion of phosphate of lime is somewhat increased, and the bone also becomes denser. This will explain why the bones of children are so elastic that, when considerable force is applied, they are apt to bend, like a green stick, rather than break, as in the adult bone under similar circumstances.

Rickets is a disease of childhood representing a diminished amount of earthy matter, when the bones become bent and distorted. The deformities are particularly marked in the bones of the lower extremity, and in pelvis and chest.

* With the exception of the teeth, bone is the hardest structure in the body.

A section of bone (Fig. 1) shows it to be composed of two kinds of tissue—*compact*, which is exceedingly dense and hard like ivory; and *cancellated* or "spongy" bone, which is a network of thin plates and columns of bone. The compact tissue constitutes the outer portion of bone, and is very much increased where great strength is required, as in the shaft of the femur or "thigh-bone." The cancellated tissue is internal, and especially abundant where bone expands to form large surfaces for support, as in the extremities of the bones of the thigh, legs, etc. Cancellated tissue is not only very light, but very strong; a portion, corresponding in size and shape to a piece of loaf or cut sugar, will support a weight of three or four hundred pounds, while a square inch of compact tissue, about one half inch in thickness, will support a weight of five thousand pounds. Bone is twice the strength of oak.

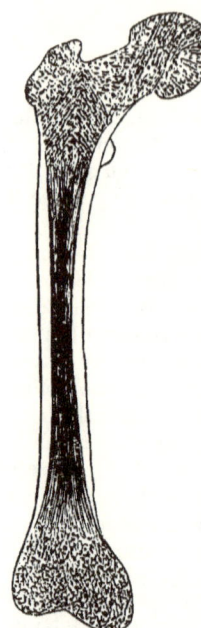

FIG. 1.—The right femur, or thigh-bone, sawn in two lengthwise. Showing arrangement of compact and cancellous tissues.

In the skull, where great strength is necessary to support and protect the brain, we find two layers of compact tissue: the external portion, known as the "outer" table, and the internal as the "inner" table, with a small amount of cancellated tissue, known as the diploë, existing between them.

The long bones are hollow. This condition admits of the proper length, diameter, and strength, while it greatly diminishes the weight.

The "medullary" or "hollow canal" in the shaft, and

the cancellated tissue of the extremities of long bones, also the cancellated tissue of other bones throughout the body, contains a substance called *marrow* or "oil of bone." In the medullary canal it is called yellow marrow, and is composed of about ninety-six per cent of fat, while the "red marrow," which is found in cancellated tissue, is three fourths water and contains but a trace of fat. The medullary canals in the bones of birds communicate with the lungs and contain air instead of marrow, thereby rendering them very light and properly adapted for flight.

Bone is supplied with nutrition from two sources; the *periosteum*, and the *nutrient arteries*. The periosteum is a firm and resisting fibrous membrane, pinkish in color, which is adherent to the bone, and covers it at all points except where cartilage exists. This membrane is a structure in which blood-vessels divide, subdivide, and pass into minute openings in the compact tissue, supplying it with nutrition. The cancellated tissue and medulla receive their nutrition from larger vessels (nutrient arteries), which are branches of the main arteries in the vicinity of the bone, and pass through the compact tissue to their destination. When the periosteum is removed from the bone, the compact tissue is deprived of its principal means of nutrition, and death or "necrosis" (corresponding to gangrene in soft tissues) is apt to ensue. This may happen in disease, or as the result of an injury, particularly to superficial bones such as the tibia, bones of the head, etc. A familiar example is a "felon," which demands immediate and special treatment at the hands of the surgeon.

When bone is in the normal condition and properly covered with periosteum, it feels smooth and moist, and when struck with a probe it gives a dull sound ; but when the periosteum is removed, it feels rough and hard, and striking it with a probe produces a metallic sound. It is important to the surgeon to recognize these different conditions, and be thus easily enabled to decide whether the periosteum is present or absent.

The *endosteum* is a thin, fragile membrane which corresponds in nutritive function to the periosteum, and lines the medullary canal in the long bones.

Bones are supplied with nerves and lymphatics or absorbents.

The *lymphatics* have been known to remove by absorption ivory pegs used to hold in place the broken ends of a bone which would not unite by the natural process.

The shape of a bone depends upon the function it performs. Bones are divided into *long, short, flat,* and *irregular*. The long bones, of which the femur or "thigh-bone" is a type, are composed of a shaft and two extremities, and, with their muscular attachments, act as levers, and also for support. The short bones are found where a number of joints are required for limited motion combined with strength. An example of this class is the carpus or wrist.

The flat bones are used in the construction of cavities and to protect their contents, and are also for muscular attachment, as the bones of the skull, the shoulder-blades (scapulæ), etc.

Examples of the irregular bones are the upper and lower jaw (superior and inferior maxillary), and the vertebræ.

The *skeleton* (Fig. 2), which represents the bones in their proper relations, is the framework to which the soft structures of the body are attached. It consists of a central column (the spine); four extremities: two upper (the arms), and two lower (the legs); and three bony cavities: (1) the cranium or skull, containing the brain; (2) the thorax or chest, containing the heart and lungs; and (3) the pelvis or basin, containing the pelvic and some of the abdominal organs.

The skeleton is composed of two hundred bones. There are several small bones found in the tendons of muscles, at the union of the skull-bones, and also in the ear, which are not included in this enumeration. This number is divided in the following manner:

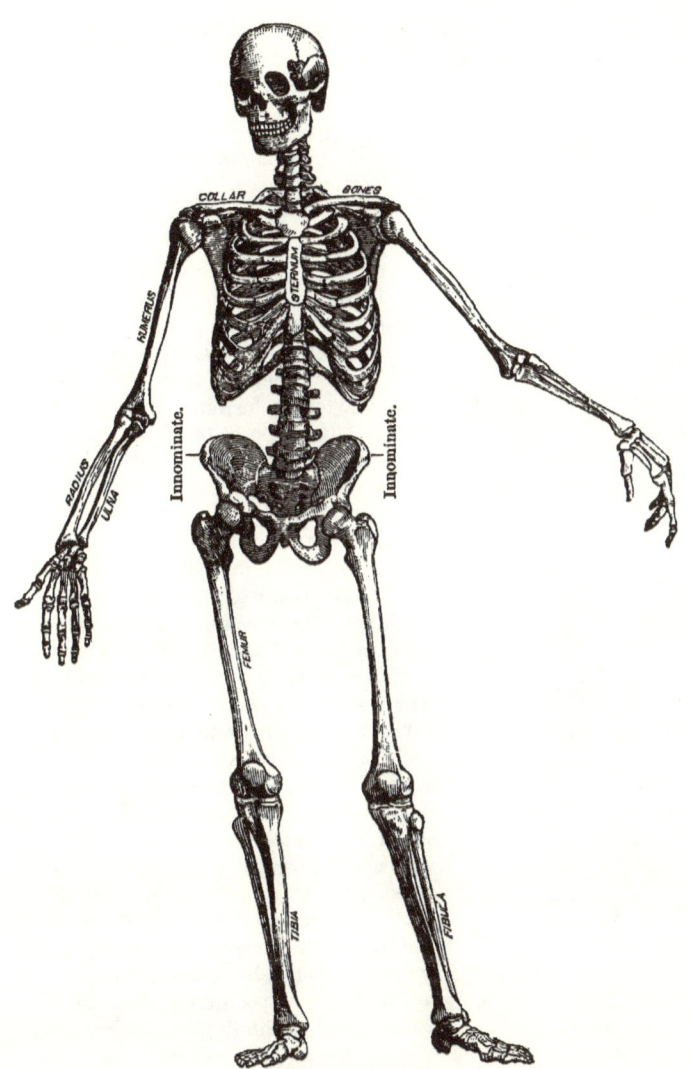

Fig. 2.—The skeleton.

The spine or vertebral column	26
Cranium	8
Face	14
Os hyoides, sternum, and ribs	26
Upper extremities	64
Lower extremities	62
	200

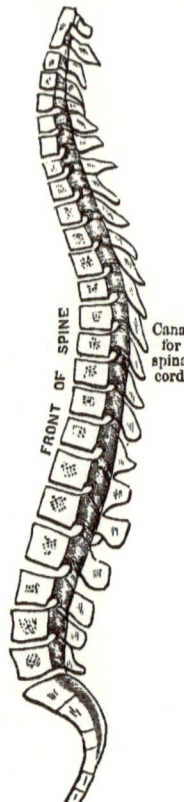

Fig. 3.—The spine, sawn in two lengthwise.

The *spine* (Fig. 3), which is about one third the length of the body, is a jointed column consisting of a series of bones (vertebræ) articulated or joined together, forming three slight curves. It supports the skull and its contents, and protects the spinal cord, which is contained in a canal formed by the union of the bony parts. At its base the column rests upon the upper portion of the pelvis—between the "haunch-bones."

A vertebra consists of two important parts—a body, in front, and a posterior portion or vertebral arch. The bodies are piled one upon the other for support and strength, the arch behind forming, by its junction with the body, a canal for the passage of the spinal cord. Between the bodies of the vertebræ are found layers or pads of a very elastic and flexible substance, known as "intervertebral cartilage," which forms a soft cushion for each vertebra to rest upon. Its elasticity prevents or diminishes shock, and also tends to restore the column to its erect position. Continued pressure on this substance during the day diminishes its thickness, consequently a person will lose about half an inch in height during this period; the rest and recumbent position of the night restore it to its previous condition.

It is important to remember that leaning too much on one side, as at a desk, will cause a permanent change in the cartilage, so that the vertebral column may be bent to one side without the bone being primarily involved.

The curves in the vertebral column are necessary to assist in forming cavities for the reception of important organs. It is said that the curves increase the strength of the column tenfold; they also aid in destroying shock.

The twenty-six bones comprising the vertebral column are divided in the following manner, viz.:

Cervical or neck............................	7
Dorsal or back..............................	12
Lumbar or loins.............................	5
The sacrum, which represents five vertebræ (in fœtal life) fused into a single bone..................	1
The coccyx or "crupper," which is formed from four bones.......................................	1
	26

The freest movement in the vertebral column is found in the neck or cervical region, and the least movement in the back, between the shoulders, the spine here being connected with the ribs. The weakest part of the spine is at the last dorsal vertebra, and this portion of the vertebral column is very movable.

It is very essential to know that different elevations on the surface of the body (bony prominences, etc.) indicate the situation of certain important internal organs; thus the vertebral column furnishes valuable information. Rubbing the fingers briskly up and down the spine will produce sufficient friction to redden the skin over the bony prominences known as "spinous processes." These processes are formed by the union posteriorly of the sides of the vertebral arches already spoken of. With the body bent forward and the arms folded across the chest, the spinous processes can be made even more apparent.

The spinous process of the seventh or lowest cervical vertebra is particularly prominent, and has received the name of "vertebra prominens," and should be carefully sought for, as it will aid in locating others.

The lower or inferior angle of the scapula or "shoulder-blade" is on a line with the seventh dorsal vertebra.

Having located the spinous processes of the different vertebræ, the following points are to be remembered, viz.:

Interval between Sixth and Seventh Cervical Spines.—Œsophagus and trachea begin.

Seventh Cervical.—Apex or upper part of lung; consequently a knife, or bullet wound, above this line, unless carried downward, would not injure the lung.

Third Dorsal.—Upper border of arch of aorta.

Fourth Dorsal.—Arch of aorta ends; upper level of heart; division of trachea into two portions; right and left bronchus.

Eighth Dorsal.—Left side, lower level of heart.

Ninth Dorsal.—Left side, lower end of œsophagus, passing through diaphragm; upper or cardiac opening of stomach and upper edge of spleen.

Tenth Dorsal.—Left side, lower level of lung.

First Lumbar.—About the middle of kidneys; lower edge of spleen.

The *skull* (Fig. 4) is the case for the lodgment and protection of the brain and its membranes, and important blood-vessels and nerves. It is composed of twenty-two bones, enumerated in the following manner:

Cranium (8 bones).—Occipital, 1; parietal, 2; frontal, 1; temporal, 2; sphenoid, 1; ethmoid, 1.

Face (14 bones).—Nasal, 2; superior maxillary, 2; lachrymal, 2; malar, 2; palate, 2; inferior turbinated, 2; vomer, 1; inferior maxillary, 1.

The cranial bones protect the brain, and are uniformly

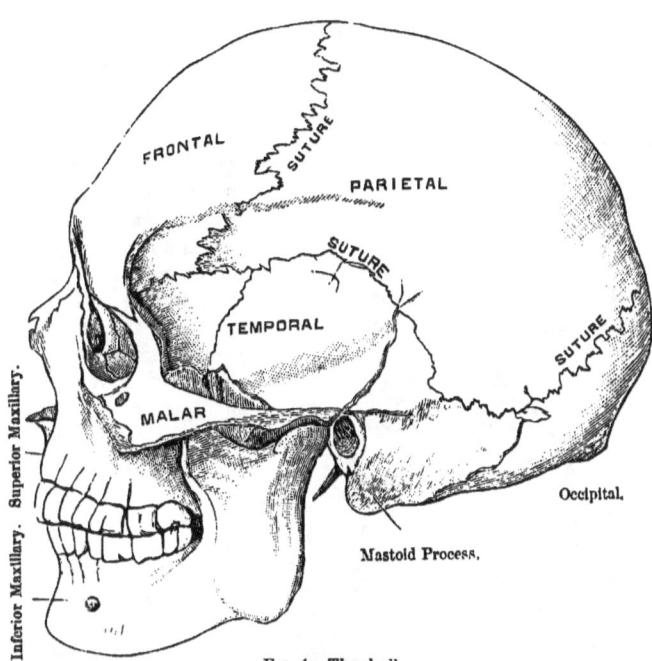

FIG. 4.—The skull.

strong and compact; while the bones of the face contain principally the organs of special sense, and give the proper symmetry to this portion of the skull, some of them are extremely thin, and are readily broken. The different bones of the skull are connected by sutures or "seams," and the overriding of these bones lessens the size of the head during birth, and permits variations, in size, for some little time afterward.

The fontanelles, or "soft spots" on an infant's head, correspond to the subsequent points of union of two or more bones of the cranium, which have not become fully hardened or ossified, but still retain the character of the membrane or soft structure from which the cranial bones are originally developed. These spots disappear when the bone becomes fully developed or ossified, which generally occurs within one and a half or two years after birth.

A large opening exists at the base of the skull in the occipital bone, called the "foramen magnum," for the transmission of the spinal cord from the brain to the canal in the spinal column. Other and smaller openings are found in the skull which transmit blood-vessels and nerves.

The skull is supported by the vertebral column. The first cervical vertebra is firmly attached to the base of the skull at the "foramen magnum," while the second cervical acts as a pivot, having a vertical or upright portion, called the "odontoid" process, which fits into the first cervical vertebra, and around which the head rotates.

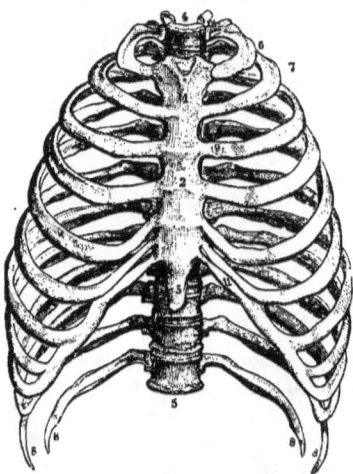

FIG. 5.—Thorax, anterior view (Sappey).
1, 2, 3, sternum; 4, circumference of the upper portion of the thorax; 5, circumference of the base of the thorax; 6, first rib; 7, second rib; 8, last two, or floating ribs; 9, costal cartilages.

The *hyoid* bone is an arch, something like a horseshoe, which is placed above the prominence on the front of the neck called the pomum Adami, or "Adam's apple," and aids in supporting the tongue, and gives attachment to certain muscles.

The *thorax* or chest (Figs. 5 and 6) is the bony frame-work which contains the heart, lungs, and important blood-vessels and nerves. It is formed

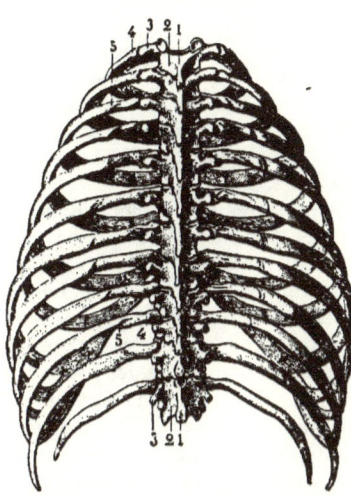

Fig. 6.—Thorax, posterior view (Sappey).
1, 1, spinous processes of the dorsal vertebræ; 2, 2, laminæ of the vertebræ; 3, 3, transverse processes; 4, 4, dorsal portions of the ribs; 5, 5, angles of the ribs.

by the sternum in front, and the ribs and vertebræ at the side and back, and is separated from the abdominal cavity by a muscular partition known as the diaphragm or "midriff."

The *sternum* or "breast-bone" is a flat and narrow bone about seven inches long, situated in the front of the chest, and supporting the clavicles or "collar-bones" and the ribs, with the exception of the last two ribs.

The *ribs* (Fig. 7) are twenty-four in number, twelve on each side, and are numbered from above downward. They articulate posteriorly with the dorsal vertebræ, and the upper ten connect in front with the sternum by means of cartilages (costal cartilages), which are interposed between the ends of the ribs and the sternum, and allow of greater motion and elasticity. The two lower ribs, the eleventh and twelfth, are simply connected with the vertebræ, and are known as "floating" ribs, as they have no attachment in front. The peculiar arrangement and attachment of the ribs render them a very important element in respiration; during this act the ribs are elevated and depressed by the action of the respiratory muscles. When the ribs are elevated during inspiration, the thorax or chest is enlarged and air passes into the lungs; in expiration the ribs are de-

12 PROMPT AID TO THE INJURED.

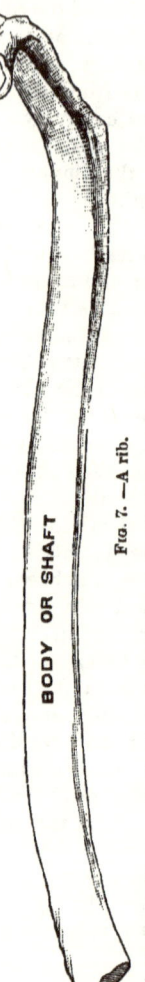

Fig. 7.—A rib.

pressed, the cavity diminished in size, and the air in the lungs is expelled. The elasticity of these bones is illustrated by the fact that children in Arabia use the ribs of camels for bows.

Upper extremities.—Each upper extremity is composed of thirty-two bones, and includes the following: scapula, clavicle, humerus, radius, ulna, eight carpal bones, five metacarpal bones, and fourteen phalanges.

The *scapulæ*, or "shoulder-blades" (Fig. 8), are thin, flat, and triangular bones, which have on their posterior and upper part an elevation or ridge known as the "spine"; the external extremity of the spine is the acromion process, which connects with the outer end of the clavicle. The outer and upper portion of the scapula forms the glenoid cavity or "socket" which receives the ball-shaped head of the humerus or "arm-bone." The scapula, in this manner, assists in the formation of the shoulder-joint. It also furnishes broad surfaces for the attachment of muscles, and aids in protecting the contents of the thorax.

The *clavicles* or "collar-bones" (Fig. 9), shaped something like the italic letter *f*, are on the anterior and upper part of the chest, and extend from the upper portion of the sternum or "breast-bone" to the acromion processes of the shoulder-blades referred to above ; they hold the shoulders upward, backward, and outward. When a clavicle is broken the

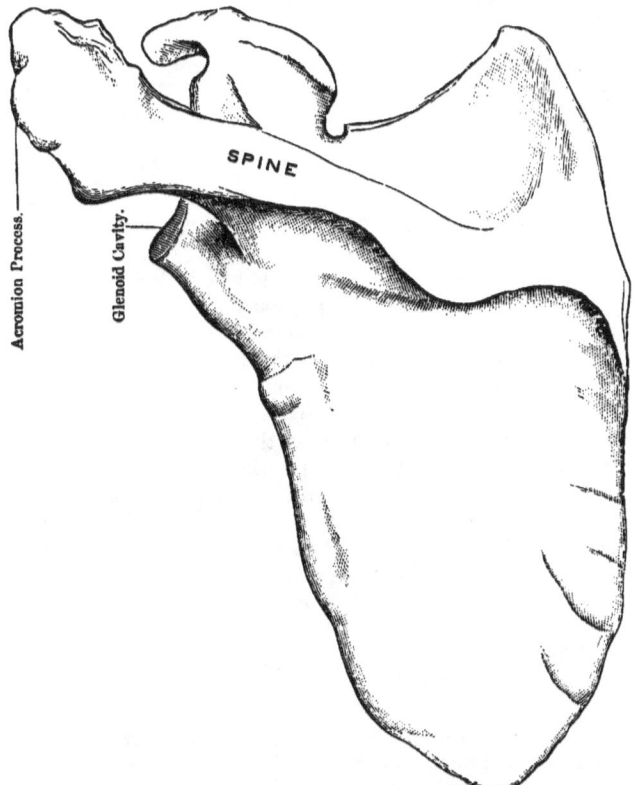

FIG. 8.—Left scapula, or shoulder-blade.

arm drops downward, forward, and inward, its support being gone. Manual labor increases the size, strength, and curvature of the clavicle. The collar-bone is less curved in women than in men.

The *humerus* or "arm-bone" (Fig. 10) is the longest and strongest bone of the upper extremity. It has a shaft and two extremities; the upper extremity is the largest part of

Fig. 9.—Anterior surface of left clavicle, or collar-bone.

the bone, and consists of a head joined to the shaft by a very short and constricted portion called the "anatomical" neck. The head is globular in form, covered with cartilage, and articulates with the glenoid cavity of the scapula, already spoken of, forming with it a ball-and-socket joint. The cartilage covering the head protects it and prevents undue friction in the joint. At the point where the anatomical neck joins the shaft are found two rough eminences called tuberosities, which are for muscular attachment. Just below the tuberosities is another constriction—the "surgical" neck—so called from the fact that it is the common seat of fracture. The shaft is partly cylindrical, prismatic, and flattened, and roughened for the attachment of muscles. The lower extremity is a broad and flattened (from before backward) portion of the bone which articulates with the ulna principally, and to a certain degree, with the radius of the forearm, these three bones forming the elbow-joint. The humerus has a greater range of motion than any other bone in the body, and is oftener dislocated; it is also frequently broken or fractured, particularly just below the head, at the surgical neck already described.

The *radius* and *ulna* (Fig. 11) are the bones that constitute the forearm.

The *radius*, so called from its fancied resemblance to

the spoke of a wheel, is the external bone lying parallel with the ulna when the palm of the hand is turned upward. It resembles other long bones in having a shaft and two extremities. The upper or lesser extremity forms only a small portion of the elbow-joint, while the lower extremity, which is broad, forms the wrist-joint by articulating with two of the carpal or wrist bones. The lower end also articulates with the corresponding portion of the ulna. The shaft of the radius is more or less irregular and rough for the origin and insertion of muscles. The ease with which the forearm and hand may be turned with the palm down ("pronation"), and with the palm up ("supination"), is due to the peculiar articulation of the upper end of the radius with the ulna. The upper end of the radius (or head) is somewhat knob-shaped, with a depression or excavation on its upper surface; the excavation articulates with a small portion of the lower extremity of the humerus, and represents the limited part it takes in the formation of the elbow-joint.

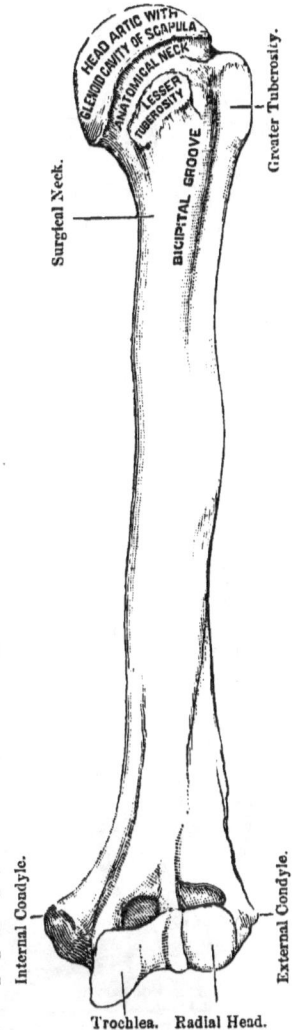

Fig. 10.—Left humerus, or arm-bone.

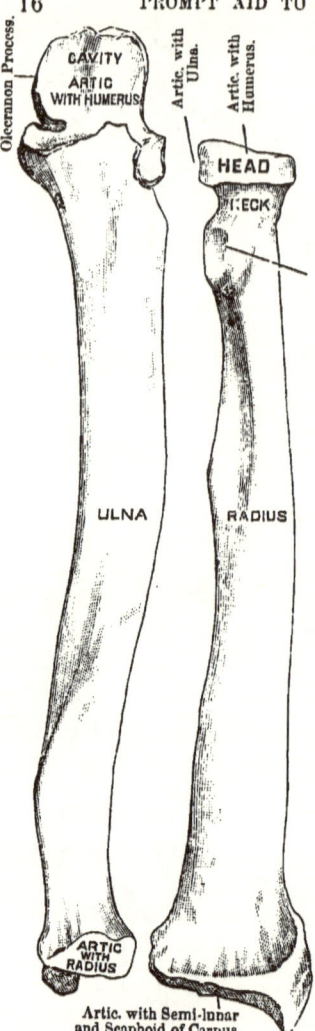

Fig. 11.—Left radius and ulna, or bones of the fore-arm.

Below the portion of the radius just described is a constricted portion, called the neck; a ligament in the form of a loop, named the orbicular ligament, is thrown around this part, and attached to the outer side of the ulna; this ligament, while it holds the radius in position, allows the neck of the radius, which it encircles, to freely rotate.

The *ulna*, which is composed of a shaft and two extremities, is the companion of the radius, and situated internal to it; the lower extremity is very small, and, although connected with the radius, has no articulation with the carpal or wrist bones, consequently it does not enter into the formation of the wrist-joint; but the upper end is large, and contains a depression or cavity which receives the lower extremity of the humerus. The prominent point of the elbow is the extreme end of the ulna, and is known as the "olecranon process" of

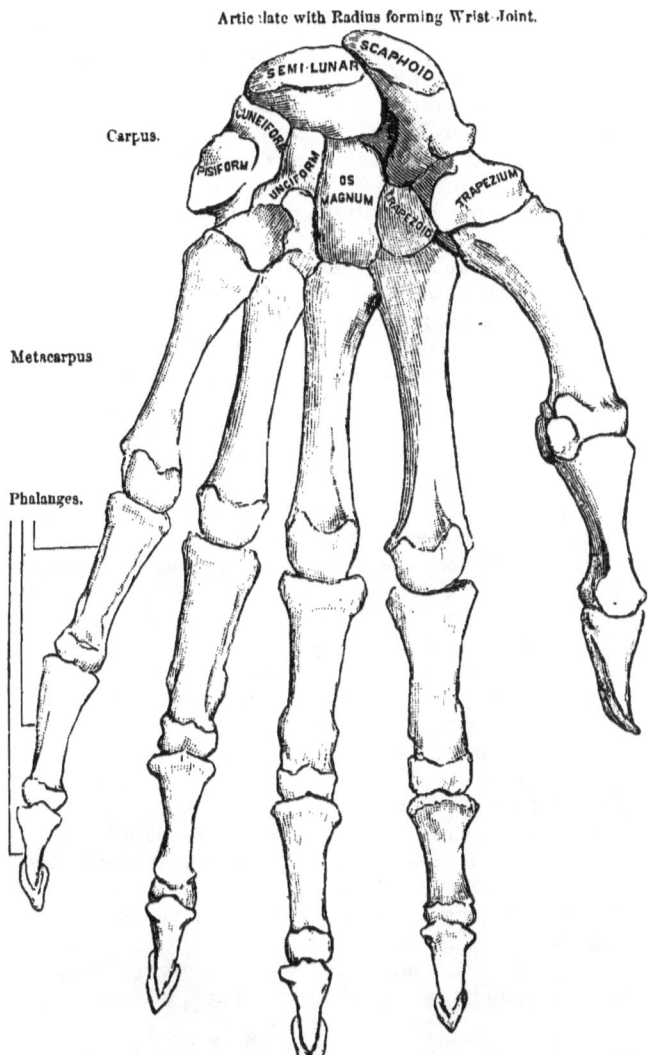

Fig. 12.—Bones of the left hand. Palmar surface.

the ulna or "funny-bone," so called from the peculiar sensation experienced when the ulnar nerve, which is closely associated with the inner border of this process, is struck. The upper extremity of the ulna and the lower extremity of the humerus make up the elbow-joint—that is, the principal part of it ; while, as already described, the upper end of the radius articulates with a small and less important part of the lower extremity of the humerus.

The *carpus* or wrist (Fig. 12) consists of the following small and irregular bones, arranged in two rows: the upper, composed of the scaphoid, semi-lunar, cuneiform, and pisiform, and the lower row, of the trapezium, trapezoid, os magnum, and unciform. The large number of bones constituting the carpus, and having so many small joints, increases its strength, motion, and elasticity. They also diminish shock and the tendency to fracture, which would be frequent if the wrist were composed of one bone. The carpus supports the hand and preserves its symmetry.

The *metacarpus* (Fig. 12) consists of five long bones (metacarpal), connecting the carpus with the bones of the fingers and thumb; they form the bony framework of the palm and back of the hand.

The *phalanges* (Fig. 12) are miniature long bones, having a shaft and two extremities, and are fourteen in number, three for each of the fingers and two for each thumb; their position can be easily demonstrated by bending the fingers. The bases of the first row of phalanges articulate with the heads of the metacarpal bones : the remaining phalanges articulate with each other. When the hand is open, the fingers do not correspond in length; but when closed, in grasping an object, there is no apparent difference. If the fingers when closed were uneven, the grasping power would be diminished. The three rows of phalanges grow progressively smaller toward the end of the fingers.

The *pelvis* or "basin" supports the trunk and protects the pelvic, and a portion of the abdominal organs. It is

BONE. 19

composed of the two innominate or "haunch" bones, the sacrum, and the coccyx; the last two have been described with the spinal column.

The *innominate* bones (Fig. 2) are situated one on each side and in front, and form the greater portion of the pelvis, while the sacrum and coccyx form the posterior portion. On the outer side of an innominate bone is found a depression or socket, known as the cotyloid cavity or acetabulum, corresponding in function to the glenoid cavity of the scapula, which receives the head of the femur or "thigh-bone."

The *femur* (Fig. 13) is the longest and strongest bone in the body; its length is characteristic of the human skeleton. In the erect position, the tips of the fingers reach to about the middle of the thigh, while in the orang-outang, the fingers reach to the ankle. This depends also on the comparative shortness of the arm in the human skeleton.

The femur is divided into a shaft and two extremities, the upper extremity being composed of a globular head,

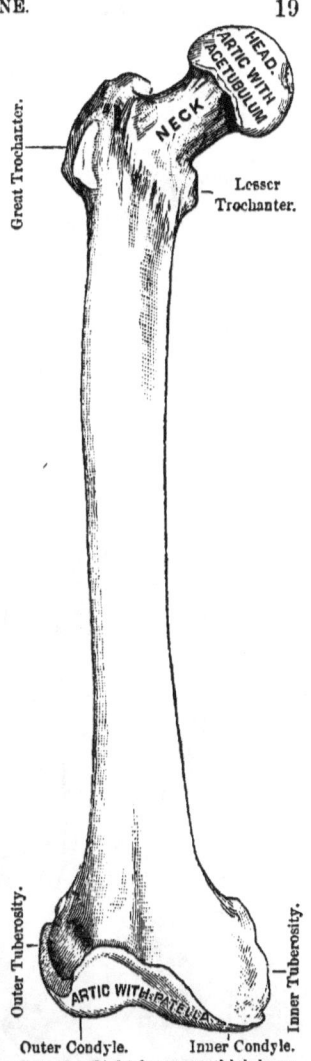

Fig. 13.—Right femur, or thigh-bone.

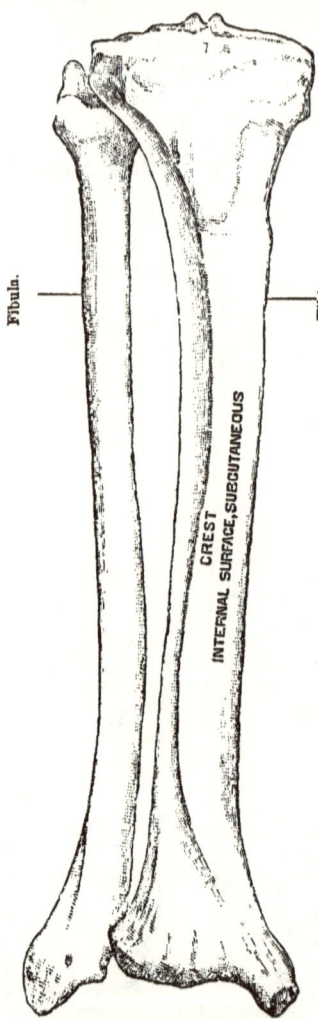

Fig. 14.—Right tibia or shin-bone, and fibula or splint-bone. Anterior surface.

which is connected obliquely with the shaft by quite a long neck; at the point where the neck joins the shaft there are two prominences, the higher being the larger, and on the outer side of the bone, and called the great trochanter, the smaller one being on the inner side, and somewhat below, known as the lesser trochanter. The eminences are for the insertion of muscles. The head of the femur is contained in the cotyloid cavity or acetabulum of the innominate or haunch bone. The long neck of the femur keeps the shaft at the proper distance from the pelvis, prevents any interference with its action, and allows greater motion. The shaft of the femur is almost cylindrical, and furnishes surfaces which are somewhat roughened for the origin and insertion of muscles. The lower extremity is very large and broad, articulates with the upper extremity of the tibia or leg-bone and the patella or knee-cap, and forms the knee-joint.

The femur articulates with the pelvis, tibia, and patella.

The *tibia* or "shin-bone" and *fibula* or "splint-bone" (Fig. 14) are the bones of the leg, and have the characteristics of long bones.

The *tibia*, which is very superficial in front (having only the skin as a covering), is the larger of the two bones, and is constructed mainly for strength, and supports the femur; it also furnishes attachment for a few muscles. The direction of the tibia is vertical.

The *fibula* is the companion-bone of the tibia and is external to it. It is about the length of the tibia, and is very slim, and helps to support the tibia and form the ankle-joint,

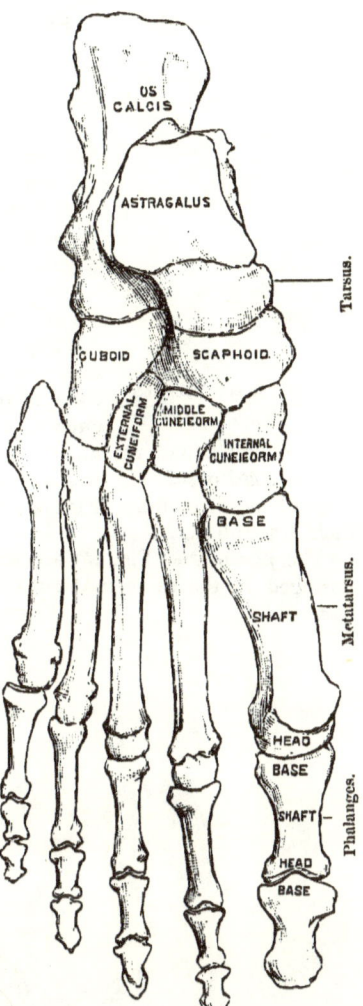

FIG. 15.—Bones of right foot. Dorsal surface.

and also furnishes the origin of a number of important muscles.

The *patella* or "knee-cap" (Fig. 16) is situated in front of the knee-joint; it is in the tendon of a muscle, and assists in protecting the joint, and also aids in properly extending the leg. When the leg is extended and the muscles are relaxed, it will be found that the patella is freely movable.

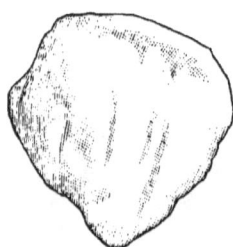

FIG. 16.—The patella, or knee-cap.

The *tarsus* or "instep" (Fig. 15) corresponds to the carpus in the upper extremity, and consists of seven bones (tarsal), viz.: calcaneum or os calcis, astragalus, cuboid, scaphoid, internal cuneiform, middle cuneiform, and external cuneiform.

The *metatarsal* (Fig. 15) bones correspond to the metacarpal bones in number (Fig. 12), and in a general way to their description.

The *phalanges* (Fig. 15) are fourteen in number, and arranged in the same manner as the phalanges of the hand.

CHAPTER II.

JOINTS—CARTILAGE—LIGAMENTS—SYNOVIAL MEMBRANE—MUSCLES.

JOINTS.

Bones are joined at different points, constituting articulations or "joints"; some are immovable, as joints of the skull; others slightly movable, as joints of the spine; while the remainder are freely movable, as the knee and shoulder joints, joints of the fingers, etc.

The movable joints are divided into four kinds:

1. Joints having a gliding movement; as the tarsus and carpus.

2. Ball-and-socket joints, as where a globular head is received into a cup-like cavity, and admitting of motion in all directions, as shoulder and hip.

3. Hinge-joints, where the motion is limited to two directions; forward and backward, as the elbow.

4. Pivot joints, where rotation only is permitted, as the joint between the atlas and axis and the joint between the radius and ulna.

Other varieties of joints sometimes described are simply modifications of the above.

Joints admit of the following movements: flexion, extension, rotation, adduction, abduction, circumduction, pronation, and supination.

Flexion takes place when the forearm is bent upon the arm, the leg upon the thigh, etc.; and *extension*, when these parts are extended or straightened; *rotation* occurs when a part is turned in and out; *adduction* is illustrated by the movement which carries the arm or leg *toward* the one of the opposite side or toward the median line of the body; *abduction*, the movement in the opposite direction; *circumduction* is exemplified when the arm is swung in a circle, as at the shoulder-joint; *pronation* is when the palm of the hand is turned downward; and *supination* is when the palm is turned upward.

The following structures enter into the formation of a joint, viz.: bone, cartilage, ligaments, tendons, synovial membrane, blood-vessels, and nerves.

CARTILAGE.

Cartilage or "gristle" is a very firm and dense substance, found principally in joints, and covers the ends of bones entering into their formation; it protects the adjacent bony surfaces against friction, and also prevents shock which would occur if the bones were directly applied to each other. The continued pressure brought to bear on cartilage in a joint would render it subject to inflammation and disease if it were supplied with blood-vessels; consequently, cartilage has no blood-supply, and is known as the non-vascular tissue; it absorbs its nutrition from the surrounding tissues by a process known as "imbibition." Cartilage connected with ribs has already been described (see RIBS). Cartilage is also found in tubes such as the air-passages, ear, nose, etc., where it is necessary that the tubes should be kept permanently open.

LIGAMENTS.

Ligaments are of two kinds, viz., the white fibrous and the yellow elastic.

The white fibrous ligaments are composed of bundles of white fibrous tissue which are closely interlaced; they are found at the movable joints, and serve to connect the extremities of the bones forming an articulation; they are inextensible but very flexible and strong, and, while they admit of the free movement necessary at the joint, they do not allow the articular ends of the bones to be abnormally displaced. When a dislocation occurs, the ligaments are either ruptured or torn from their attachments. The ligaments composed of yellow elastic tissue are extremely extensible, and are fewer in number than those composed of white fibrous tissue. Examples of this variety are the ligamenta subflava and the ligamentum nuchæ. The latter extends from the spinous process of the seventh cervical vertebra to a protuberance on the occipital bone at the base of the skull. It is rudimentary in man, but well developed in animals, where it aids in supporting the head, and acts as a substitute for muscular power. The "ligamenta subflava" are small ligaments of the same tissue (yellow elastic) connected with the different vertebræ composing the spinal column; their action is similar to that of the ligamentum nuchæ, and assist in bringing the vertebral column to an erect position.

SYNOVIAL MEMBRANE.

The synovial membrane is a delicate structure connected with movable joints and covering the inner side of the ligaments. It is not found between the articular surfaces of the bones composing the joints. The synovial membrane secretes a thickish, glairy fluid, resembling in appearance the white of an egg, and it is known as synovia or "joint oil," which, passing into the joints, lubricates the cartilages and prevents friction. The synovial membrane may become inflamed as the result of injury or disease, producing stiffness of the joint, or anchylosis.

MUSCLES.

Muscles are bundles of reddish-brown fibers having the power of contraction, and are divided into two kinds, viz., voluntary or striped, and involuntary or unstriped; the terms striped and unstriped refer to the microscopical appearance of the fibers composing the muscle.

The *voluntary* muscles are under the control of the will, and comprise the bulk of the muscular system; they are attached to the different parts of the skeleton, and act on the bones as levers; they are the agents of locomotion and of the movements of all portions of the body; they also protect the different structures beneath them, and give

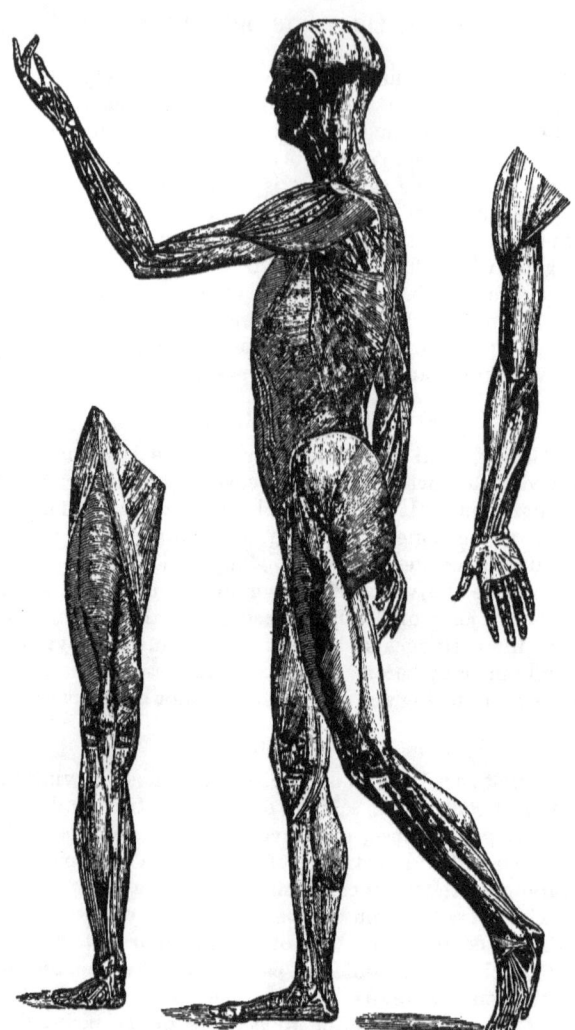

Fig. 17.—The muscular system.

grace and symmetry to the form (Fig. 17). When muscles become diminished in size or illy developed, the person is angular or bony.

The lean meat used for food is muscular tissue.

Muscles are of different lengths and shapes—long, short, broad, etc. They are arranged in pairs—that is, there are corresponding muscles on each side of the body, although some exist singly. There are over four hundred muscles in the human frame. In many situations the outlines of muscles are apparent to the eye, and are utilized as guides to the location of important blood-vessels and nerves.

Muscles are surrounded by a thin, web-like tissue known as fascia, which serves as a support for them and allows of their movements without undue friction.

Muscles are composed of a belly and two extremities. The more fixed extremity is called the origin, and the movable one the insertion; however, in some muscles the origin and insertion are equally movable. During contraction their origins and insertions approximate, and muscles shorten and thicken; this can be illustrated by raising a heavy weight with the hand, when the biceps muscle in the arm will be seen to thicken and bulge forward. Muscles gradually diminish in size toward their extremities, the muscular tissue being replaced by an extremely firm and resisting substance called white fibrous tissue, constituting a *tendon*.

The tendons are directly and intimately connected with the periosteum covering the bone, and sometimes directly with the bone. They differ from muscular tissue in appearance—being white and glistening.

An *aponeurosis* is simply an expanded tendon, and is found where the muscle has a broad attachment.

Muscles are abundantly supplied with blood-vessels, nerves, and lymphatics, but tendons have few blood-vessels and only those of the largest size have nerves.

The nerves connected with the muscles transmit to them a stimulus from the brain and spinal cord, which calls their function into action. In the disease known as tetanus, or "locked-jaw," the stimulus is sometimes so great, and the contraction so intense, that the muscular fibers are ruptured.

The stiffness that ensues after death (rigor mortis) is due to a solidification of some of the fluid portions of the muscle, which is succeeded after a short time, however, by relaxation of the body. It is said that this rigidity is not present, or only very transient, in one killed by lightning.

If a muscle is not sufficiently used it becomes diminished in size (atrophy), and from continued non-use may undergo degeneration, from which there is no recovery ; consequently a muscle should not be confined or unused too long. Hypertrophy represents the opposite condition to atrophy, the muscles being increased in size, and may be the result of constant exercise ; it is often noticeable in athletes. The contraction of a voluntary muscle is very rapid and abrupt.

The *involuntary* or unstriped muscles are not connected with bones, but form the muscular portion of the stomach, intestines, and other internal organs, and also the muscular coat of the blood-vessels. The contraction of this form of muscular fiber is slow, unequal, and does not affect all portions of the muscle simultaneously. The involuntary muscles are not, as a rule, under the control of the will ; they have no tendons ; the muscular fibers simply interlace with each other. Although the muscular fibers of the heart are striped, like *voluntary* muscular fibers, this organ is an *involuntary* muscle, and not under the control of the will, this being the only exception to the rule.

Reference will frequently be made in the following chapters to organs, glands, mucous and serous membranes, and also to secretion and excretion, so then for the sake of convenience these terms will now be explained.

An organ is a part of the animal organism having a

special function to perform, as the brain, heart, stomach, kidneys, etc.

A gland is an organ, but it also has the additional function of abstracting from the blood, material which it discharges from the body unchanged (excretion), as urine, or, it manufactures from certain parts of the blood, a fluid (secretion) which has a particular function, as the gastric juice.

Mucous and serous membranes line the interior of cavities, sacs, tubes, etc. *Serous membranes*, except the peritonæum in the female, line cavities and tubes which are closed and have no communication with the outer world (examples: pericardium pleura, and peritonæum); while *mucous membranes* line structures which open externally (examples: respiratory and alimentary tracts). Both membranes secrete a small amount of fluid, which keeps their surface moist and pliable, and prevents friction; in addition to this, some mucous membranes have small glands in their structure which produce a secretion having a distinctive function, as the mucous membrane lining the stomach, which secretes the gastric juice.

A *secretion* is a fluid formed in and by a gland or organ for a special purpose. It is peculiar to the organ that produces it and is found in no other part, and consequently never exists in the blood; examples: gastric juice, bile, etc.

An *excretion* represents certain material which is always present in the blood, composed principally of waste matter; it is not formed in the excretory organ, but is simply removed from the blood and discharged from the body. The urine is an example of an excretion, the kidneys being excretory organs.

Some organs have both excreting and secreting functions; the liver is an example of this type.

CHAPTER III.

THE BLOOD AND CIRCULATORY ORGANS.

BLOOD is the great nutritive fluid of the body. It distributes to the different tissues material necessary for their proper maintenance and activity. It supplies heat and also oxygen, receiving the latter from the air. It is absolutely necessary that the tissues should be constantly furnished with oxygen, otherwise death would ensue from suffocation, as in drowning. The blood has also the important function of removing from the body through certain organs, principally the kidneys, lungs and skin, worn-out and waste matter, which, if retained in the body, would produce serious disease or death.

Blood is composed of a liquid and a solid portion, about equally divided. The liquid portion or "plasma" is almost colorless when separated from the solid matter or "corpuscles," and contains the principal elements of nutrition, which it distributes to the different tissues in its course throughout the system, receiving in return waste material which is to be discharged from the body.

The corpuscles or solid portion float in the plasma, and constitute about one half of the bulk of the blood. They are divided into three kinds, viz., red corpuscles, leucocytes or white corpuscles, and those known as blood plaques. The red corpuscles are by far the most important and numerous; they resemble in appearance a thick coin which has been made thin in the center on both sides, or biconcave. These little bodies are exceedingly small; thirty-five hundred of them laid side by side on their flat surface would be required to cover the space of one inch. They are, of course, not visible to the naked eye. When examined by the microscope they appear straw-colored; but when vast numbers are crowded together, as in the normal condition of the blood, they give to it the characteristic red color. Physiologists have calculated that there are over fifty billions of them in the human body. The red corpuscles are the agents that receive oxygen from the air in the lungs and distribute it to the tissues throughout the body. The white corpuscles are larger than the red (about $\frac{1}{2500}$ of an inch in diameter), and are globular in form. The blood plaques are exceedingly small, only about $\frac{1}{17000}$ to $\frac{1}{10000}$ of an inch in diameter; both the plaques and the white corpuscles are much fewer than the red corpuscles, and their function is unknown (Fig. 18).

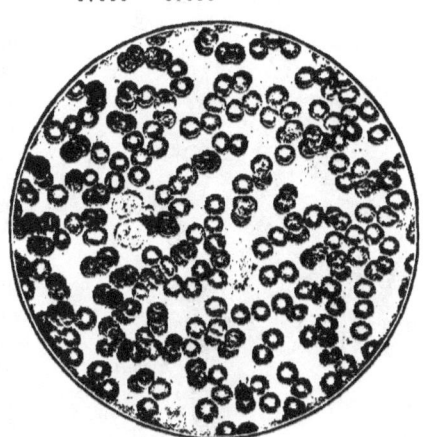

Fig. 18.—Human red-blood corpuscles, and two white corpuscles.

The amount of blood in the human body represents about one twelfth to one thirteenth of its weight. It is an alkaline fluid, and somewhat heavier than water. Arterial blood, or that which has been purified in the lungs, is of a bright red color, while the venous blood is purplish.

Some tissues, such as the hair, cartilage, nails, etc., are not supplied with blood, but receive their nutrition by absorption (imbibition) from surrounding parts.

Coagulation, which occurs when a blood-vessel is opened, is an effort of Nature to arrest hæmorrhage. It also takes place in the blood-vessels in certain abnormal conditions of the blood. Blood circulates throughout the body by means of blood-vessels, which are divided into arteries, veins, and capillaries. The motive power or force that propels the blood through these vessels is furnished by contractions of the heart.

THE HEART.

The *heart*, which is likened to a pump, is a hollow, muscular organ, conical or pear-shaped, about the size of a closed fist, four or five inches long, three inches through, and weighs about eight to twelve ounces (Fig. 19). It occupies a position in the thorax or chest, just behind the sternum or "breast-bone," and between the lungs, but mainly on the left side. The large end or "base" of the heart is directed upward and toward the right side, while the smaller end or "apex" is downward and to the left. A line drawn across the sternum at the upper border of the cartilage forming the extremity of the third rib would indicate the base of the heart. The apex is found in the space between the fifth and sixth ribs, three and a half

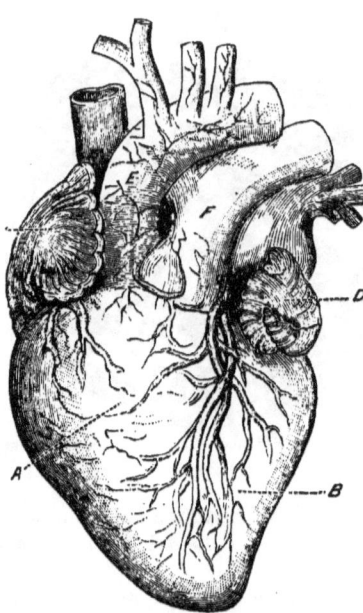

FIG. 19.—The heart and large blood-vessels. *A*, right ventricle; *B*, left ventricle; *C*, right auricle; *D*, left auricle; *E*, aorta; *F*, pulmonary artery.

inches to the left of the middle line. At this point the pulsation of the heart may be distinctly felt (Fig. 20).

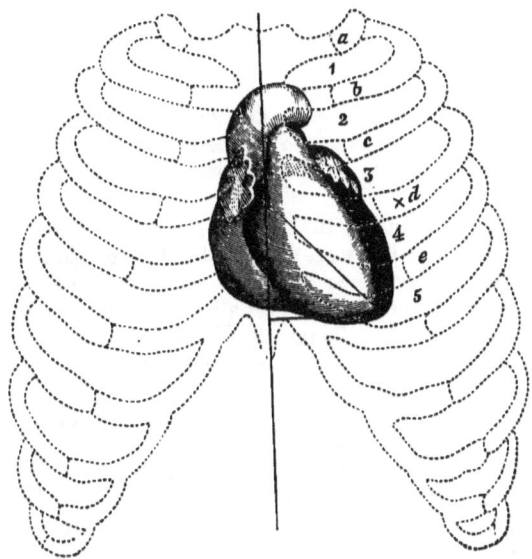

FIG. 20.—Heart and ribs. *a, b, c, d, e,* ribs; 1, 2, 3, 4, 5, intercostal spaces; *x,* position of nipple (fourth rib).

The heart contains four cavities or chambers—two auricles and two ventricles; the auricles are above and the ventricles are below, or, the heart may be divided into a right and left side, each having an auricle and a ventricle. The ventricles do not connect with each other, nor do the auricles, but an auricle connects with the ventricle of the corresponding side. The openings between these cavities are guarded by the auriculo-ventricular valves, and they are so arranged that the blood may pass from the auricles to the ventricles; but it can not in the normal condition return. If this were possible, the circulation of blood would be constantly interfered with. The ventricles are larger than the auricles and their walls are stronger, particularly the walls of the left ventricle, which drives the blood received from the lungs throughout the entire system. The auricles contract simultaneously; the contraction of the ventricles follows immediately afterward. Between the end of the contraction of the ventricles and the beginning of the contraction of the auricles is a short pause. This double action represents a pulsation of the heart, of which there are from sixty to eighty during the minute.

The heart and also the beginning of the great vessels at the base of the heart are surrounded by a closed sac (the pericardium) composed of fibrous tissue and lined with a serous membrane. The inner wall or visceral layer of this serous structure is adherent to the heart; between it and the outer or parietal layer exists more or less of a space, being better marked at the lower portion. The serous membrane lining the space secretes a fluid, of which there is usually about one or two teaspoonfuls present, thus keeping the membrane soft and moist, and limiting friction when the two surfaces rub against each other during the action of the heart. The broadest portion or base of the pericardium, which is the outer layer, corresponds in situation to the apex of the heart. This arrangement prevents any interference with the movement of the apex of the heart, which would ensue if the apex or smaller portion of the pericardium were below. The base of the pericardium is attached to a portion of the upper surface of the diaphragm or "midriff."

The circulation of the blood in the system is effected in the following manner: The venous blood throughout the body is collected by two large veins—the superior and inferior vena cava; the superior vena cava receiving the blood from the head, upper extremities, and parts above the diaphragm, while the inferior vena cava collects the venous blood from those parts below the diaphragm. These vessels are connected with and discharge their contents into the right auricle, which then contracts, and the blood is forced into the right ventricle through the right auriculo-ventricular opening. The contraction of the ventricle which follows, closes the valves guarding this opening; the blood, being then unable to return to the auricle, is forced out of the ventricle into the pulmonary artery, a large blood-vessel connected with the right ventricle; valves also guard the opening into this vessel. The pulmonary artery divides into two branches, carrying the blood to each lung; after reaching these organs, the branches of the artery grow smaller and exceedingly numerous, and at last they become minute vessels known as capillaries, which surround the air-vesicles of the lungs. The venous blood has now been carried from the heart to the lungs, and it is at this point that the blood in the capillaries surrounding the air-vesicles frees itself of carbonic-acid gas and some other impurities, and receives in return oxygen from the air contained in the vesicles. The blood is now changed in character: the color, instead of being blue or venous, is now red or arterial, and thus enriched and purified, and in the condition to nourish the tissues, is carried from the lungs to the left auricle of the heart by four large blood-vessels, the pulmonary veins. These vessels are continuous with the pulmonary artery through the medium of the capillaries which surround the air-vesicles, and have been already referred to. The left auricle, after receiving the blood from the pulmonary veins, contracts and its contents pass into the left ventricle, which, being filled, immediately contracts and forces the blood into the aorta and closes the left auriculo-ventricular valve, thus preventing the return of the blood to the auricle. The opening from the ventricle into the aorta is also guarded by valves which bar the return of the blood to the ventricles. The aorta begins at the left ventricle, is the main artery of the body, and through it passes the arterial blood into the smaller arteries and capillaries throughout the system. In these latter vessels the nutrition contained in the blood is given direct to the tissues, and the blood receives in exchange the waste elements which are to be discharged from the body. After performing this function, the blood passes from the capillaries into the veins, the latter discharging their contents into the right auricle through the superior and inferior vena cava, as already described; the blood has then made the circuit of the body, or, in other words, the circulation is completed.

The heart is endowed with enormous power. The strength developed in twenty-four hours would raise a ton-weight about one hundred and twenty-five feet from the ground.

A very smooth and delicate membrane (the endocardium) lines the heart, and is continuous throughout the vascular system, forming the internal coat or lining (endothelium) of the arteries and veins, and is the only coat of the capillaries.

THE BLOOD-VESSELS.

The *blood-vessels*, by which the blood is carried throughout the system in response to the contraction of the heart, are divided into *arteries, capillaries*, and *veins*.

Arteries.—Arteries are the vessels that carry blood *from* the heart. The aorta (the largest artery in the body), which is the beginning of the arterial system, is connected with the left ventricle, the contents of which pass through the aorta into the smaller arteries, and thence into the system.

The walls of arteries are composed of three coats or layers, the external, the middle, and the internal. The external coat consists of white fibrous tissue, which is very strong and tough, and protects the vessel. The resistance of the fibrous coat is well illustrated in the phenomena that follow the ligation, or tying, of an artery. The ligature, which is passed around the vessel and tightly drawn together and tied, severs the middle and internal but not the external coat. The middle coat is composed of muscular and yellow elastic tissues, which exert a normal mean pressure upon the contents of the vessels, and regulate the blood-supply to the different parts of the body. The yellow elastic tissue is the principal cause of the greater thickness of the walls of the arteries compared with veins; it also accounts for the elasticity which is characteristic of arteries, and the fact that arteries remain open when empty. The latter condition gave rise, years ago, to the belief that these vessels contained air, and they were accordingly named arteries, or "air-carriers."

Arteries pulsate, and, when opened, the blood escapes in spurts or jets, the color being scarlet.

Capillaries.—When arteries become very small and lose their external and middle layers, they consist of but one coat, and are known as capillaries. These vessels are often so minute that the red blood-corpuscles, in order to pass through them, are obliged to "double up" and pass through one by one. The thinness of the capillary walls, and the slow and uniform current of blood in these ves-

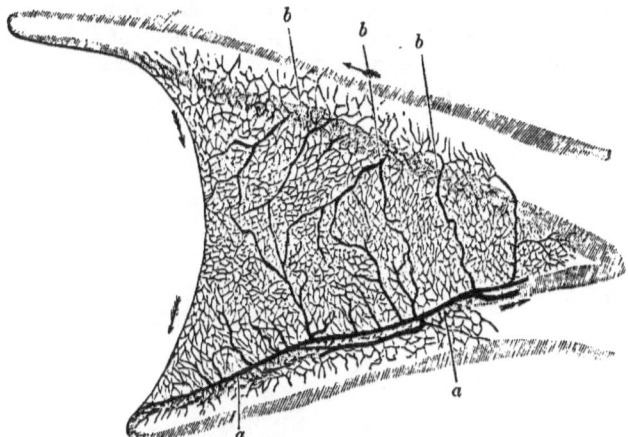

FIG. 21.—Web of frog's foot, magnified. *a, a,* veins ; *b, b,* arteries, with capillaries between them.

sels, enable them, as has already been described, to give direct to the tissues the elements of nutrition which they contain, and receive in exchange the waste of the body (Fig. 21).

Veins.—Veins carry blood *toward* the heart. These vessels, like arteries, are composed of three coats, the external and internal being similar to arteries. The middle coat, however, contains but very little yellow elastic tissue, which accounts for the comparative thinness of the walls of veins, and also the fact that they collapse when empty.

Veins, with the exception of those in the cranial, thoracic, and abdominal cavities, are supplied with valves, which are formed by a duplication of the internal coat, and allow the blood to flow in but one direction—toward the heart. This is essential, as the current of blood in veins is mainly upward, toward the heart and against the force of gravity; and also as the pressure of blood in the veins is only about one fourth that of arteries. Sometimes the valves are rendered useless, and distention and distortion of the vessel occur, as in varicose veins of the leg. This condition is associated with more or less prominence and deformity of the vein.

CHAPTER IV.

RESPIRATORY APPARATUS AND RESPIRATION.

RESPIRATION is the function which provides for the proper entrance of oxygen into the blood, and the exit from the blood of carbonic acid and certain waste matter. The respiratory apparatus for the transmission of air to the lungs includes the following structures: the mouth and nose, larynx, trachea, bronchial tubes, and air vesicles or cells, the lungs being chiefly composed of the latter. An examination of the throat will show two openings, the back or posterior one being the upper part of the œsophagus or "gullet"; in front of this, and just behind and below the base of the tongue, is the upper part of the windpipe, known as the larynx (Fig. 22).

The *larynx* is a cartilaginous box, containing fibrous bands, stretching from before backward, two on each side, superior and inferior, and called vocal cords. The upper or superior pair have no special function that is at present known. The lower or inferior cords are extremely important; the vibration of these during expiration produces the voice. The larynx may be easily located; externally the prominence known as the "Adam's apple" forms the upper portion, the lower border being about one inch and a half below. The opening into the larynx is covered by a leaf-shaped piece of cartilage, known as the epiglottis, which prevents food and other substances from entering the windpipe.

The *trachea* or "wind-pipe" is the continuation downward from the larynx of a tube about four or five inches

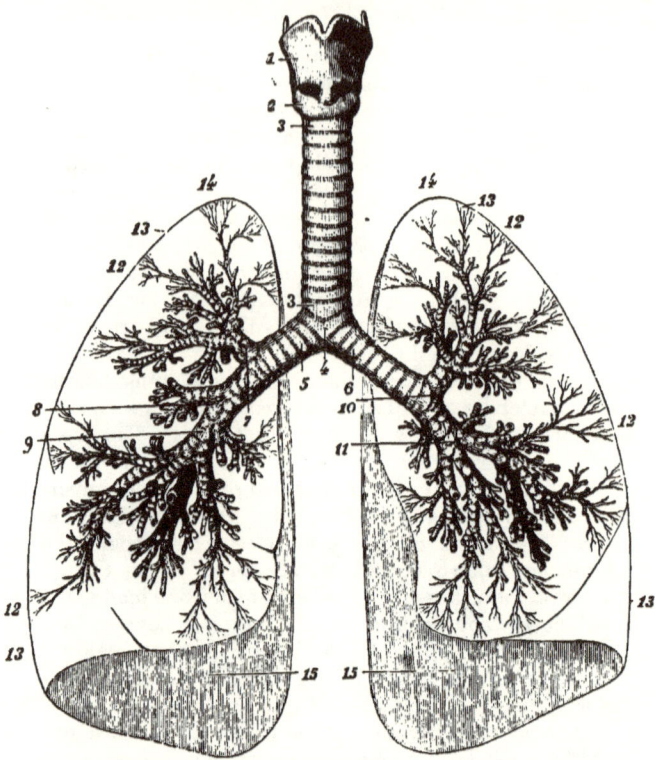

Fig. 22.—Trachea and bronchial tubes (Sappey). 1, 2, larynx; 3, 3, trachea; 4, bifurcation of the trachea; 5, right bronchus; 6, left bronchus; 7, bronchial division to the upper lobe of the right lung; 8, division to the middle lobe; 9, division to the lower lobe; 10, division to the upper lobe of the left lung; 11, division to the lower lobe; 12, 12, 12, 12, ultimate ramifications of the bronchi; 13, 13, 13, 13, lungs, represented in contour; 14, 14, summit of the lungs; 15, 15, base of the lungs.

long and three quarters of an inch in diameter, composed of cartilaginous rings, fibrous membrane, and a small amount of muscular tissue. These rings have the same use as the cartilage forming the larynx—to keep their respective walls separated at all times. The trachea begins opposite the interval between the spinous processes of the sixth and seventh cervical vertebræ and ends at the level of the spinous process of the fourth dorsal vertebra, where it divides into two branches, the right and left bronchus. These tubes after entering the lungs divide into a great number of branches, or bronchial tubes, which further divide and subdivide until they become exceedingly minute and end in little pouches known as air-cells or vesicles, which are very small and numerous and will be again referred to in describing the lungs. The respiratory tract is lined throughout by a mucous membrane, which is kept moist and prevents friction during the passage of the air through the tube. Its construction is peculiar,

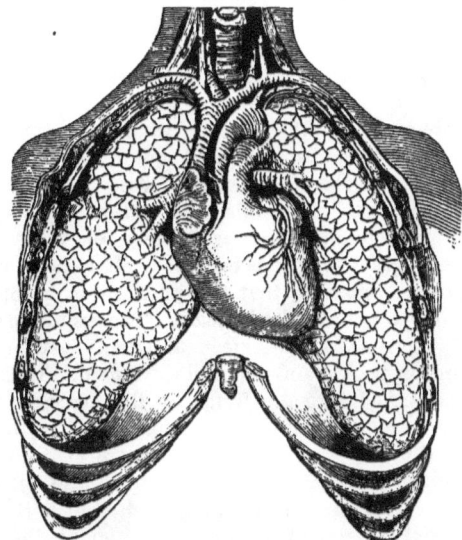

FIG. 23.—Showing the relative position of heart and lungs in the cavity of the chest.

having little hair-like processes which are constantly waving toward the outer world, assisting in preventing the entrance into the lungs of dust and irritating particles.

The *lungs* (Figs. 22 and 23) are two pyramidal or cone-shaped organs, longer in the back than in the front, situated in the chest, each weighing about twenty ounces, the right one being a little heavier than the left. They are divided by deep fissures into lobes, the right lung having three and the left two lobes. The small or upper portion or apex, extends to or just above the clavicle or collarbone; the larger or lower portion or base, descends in front to about the sixth rib, on the side to the eighth, and in the back to the tenth. Each lung is covered by a closed sac called the *pleura*, one of its layers being adherent to the lung and the other lining the chest-walls. The cavity of the pleura contains a small amount of fluid, to prevent friction during the action of the lungs. The lungs are composed of millions of air-cells or pouches, about one two-hundredth of an inch in diameter, which are the termination of the bronchial tubes (Fig. 24). The enormous space represented by the air-cells would, if spread out, cover an area of about six hundred square feet. This will give some idea of the vast surface exposed to the air passing into the lungs. The walls of the air-cells and of the capillaries surrounding them are so very thin that the interchange of gases readily takes place. It is here, as has already been explained, that the purification or change from venous to arterial blood occurs. The respiratory act, of which there are from sixteen to twenty per minute, consists of inspiration and expiration. During inspiration the air is carried into the lungs by the descent of the diaphragm, which exerts a suction-force, and by the elevation of the ribs, which increases the size of the chest. Expiration, or the expulsion of air from the lungs, is effected by elevation of the diaphragm, descent of the ribs, and a partial collapse of the lungs. Although these are the principal agents of inspiration and expiration, there

RESPIRATORY APPARATUS AND RESPIRATION. 43

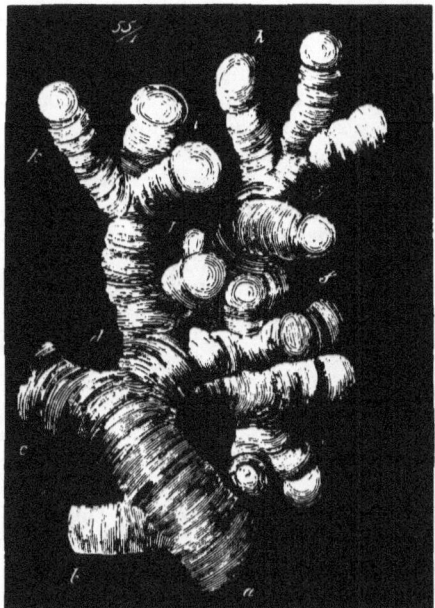

FIG. 24.—Terminal bronchial tubes ending in air-cells.

are other elements, notably the assistance of certain muscles, which, although of lesser importance, aid this function; and, when there is difficulty in breathing, as in some diseases of the lungs, the action of the auxiliary muscles about the neck and shoulders becomes very apparent.

The lungs during life are never entirely collapsed, this being prevented by about one hundred cubic inches of air which can not be expelled, and is called "residual air." Another hundred cubic inches of air, known as "reserve air," usually remains in the lungs after expiration, and is used by these organs during any increased physical exertion, as in running, etc., which requires an extra amount of air.

The "tidal air" represents the amount of air taken into the lungs at each ordinary inspiration, and consists of thirty cubic inches. During violent exercise, however, an additional one hundred cubic inches is taken into the lungs at each inspiration, and is known as "complemental air." The extreme capacity of the lungs would, consequently, be the sum of the residual (100), reserve (100), tidal (30), and complemental (100) volumes of air, amounting to three hundred and thirty cubic inches. The "vital capacity" or "respiratory capacity," however, is the amount which can be breathed out after the deepest possible inspiration, and would include, therefore,

   Complemental air - - - 100 cubic inches.
   Tidal air - - - - - 30 " "
   Reserve air - - - - 100 " "
    Total - . - - - - 230 cubic inches.

The air which we breathe is composed of two gases—oxygen (21 parts) and nitrogen (79 parts). There are also, in varying quantities, vapors, traces of ammonia, etc., the amount depending on location, It is the addition of poisonous elements to the air, as in large cities, or where bad sanitary conditions exist, that furnish the causes of certain diseases.

CHAPTER V.

ALIMENTATION AND DIGESTION.

THE alimentary tract or canal comprises the several structures or organs through which food and drink pass to be properly digested and absorbed. That portion of the food or material which is not needed for nutrition, or can not be acted upon by the different secretions, is discharged from the body. The alimentary canal, which is about thirty feet long, begins at the mouth; then follows the pharynx or "throat;" below it is the œsophagus or "gullet," then the stomach, then the small and large intestine, the canal terminating at the lower end of the latter, at the external opening called the anus, the "fundament," or "body."

These structures will be spoken of separately in their order from above downward. Other organs upon which the function of digestion is dependent, and which are connected with and discharge their secretions into the alimentary tract, will be described in their proper connection.

Mastication is the first step in digestion. This takes place in the mouth, and is performed by the teeth, which are so fashioned and arranged that the food may be cut, torn, and ground, showing that man is adapted to all kinds of food. In animals, such as the cow, that secure their food principally by grazing, the molars or "grinders" are particularly well developed, while in dogs and animals that depend principally upon meat for sustenance, the sharp or tusk-like teeth called canines are very prominent, and enable them to tear the meat from the bone. The teeth are inserted in the superior and inferior maxillæ, or upper and lower "jaw-bones," along their edge, or alveolar process. There are thirty-two teeth in all—sixteen in each jaw—and arranged in the manner shown in the diagram (Fig. 25). That portion of a tooth projecting beyond the gums is called the crown; the root is imbedded in the bone. The crown is covered with enamel, the hardest structure in the body, which protects the teeth and prevents them from wearing out as the result of the friction during mastication. Acids, if used too freely or too strong, are very destructive to the enamel, dissolving out the lime of which this structure is composed. The use of acids is often indicated for medical treatment; in this case, immediately after the acid is taken the mouth should be washed out with a solution of bicarbonate of soda (common baking soda), which neutralizes the acid. The teeth cut up and grind the food so that all parts of it may be exposed to the different digestive secretions. Improper mastication, or an absence of teeth, will be followed by imperfect digestion. As mastication progresses, the

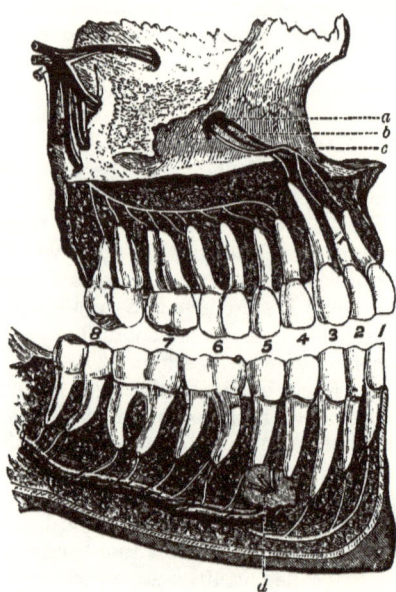

FIG. 25.—The jaws and the teeth. 1, 2, incisors; 3, canines; 4, 5, bicuspids; 6, 7, 8, molars; *a*, vein; *b*, artery; *c*, nerve; *d*, vein, artery, and nerve.

food is made soft and wet by an alkaline secretion known as saliva or "spittle," which is secreted by three salivary glands—*parotid, submaxillary,* and *sublingual.* These open by small ducts into the mouth (Fig. 26). The parotid is the largest one, and is situated behind the angle of the jaw. The characteristic deformity in the disease known as "mumps" is the enlargement of this gland. The secretion of the parotid is used mainly to moisten the food, while that of the submaxillary and sublingual is more viscid or slippery, coating and allowing it to pass down the œsophagus to the stomach with great ease. In reptiles, where there is no mastication, it is this slimy secretion that enables them to swallow substances of very large bulk. Saliva also acts chemically on food and changes a part of the starchy matter into sugar. During mastication the cheeks, lips, and tongue keep the food between the teeth.

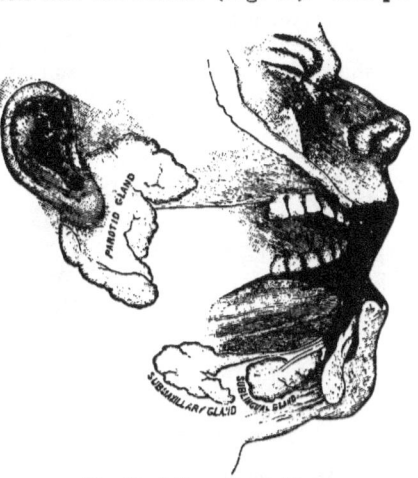

FIG. 26.—Salivary glands (Tracy).

The *pharynx,* or throat, is the continuation of the mouth, and has no special digestive action while the food is in this portion of the tract.

The *œsophagus,* or gullet, the next portion, is a muscular tube about eight or nine inches long, and is collapsed when not in use. It begins at the lower border of the pharynx, opposite the interval between the spines of the sixth and seventh cervical vertebræ, passes downward through an opening in the diaphragm, and becomes continuous with the cardiac end of the stomach opposite the ninth dorsal vertebra. This tube by its muscular action accelerates the passage of food downward to the stomach.

The *stomach* (Fig. 27) is one of the principal organs of digestion. It appears to be a dilated portion of the alimentary canal. In shape it somewhat resembles a bag-pipe, having a greater and lesser curvature it lies crosswise in the abdominal cavity, and has two openings, one on the left side, which is continuous with the lower end of the œsophagus in the vicinity of the heart, and called the cardiac opening, the other or pyloric opening being on the right side ; this opening is the beginning of the small intestine, and is guarded by a valve-like constriction, the pylorus, or " gate-keeper."

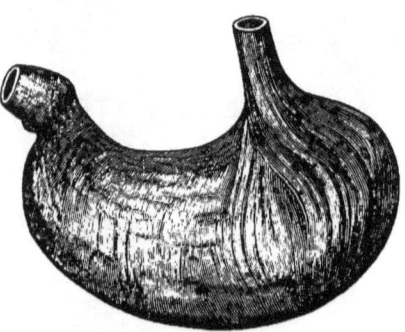

Fig. 27.—The stomach.

The stomach is about fourteen inches long, five inches in diameter, and weighs four ounces; its capacity being about five pints. It is composed of three coats or layers: the external, fibro-serous; the middle, muscular; and the internal, mucous. The external coat prevents friction during the movement of the stomach and also adds strength. The muscular portion produces the movements necessary for the proper digestion of the contents of the stomach. The internal coat or mucous membrane is the most important layer. In the substance of this membrane are found numerous little glands which open upon the surface of the membrane. These peptic or gastric glands, secrete the gastric juice, which is so necessary to digestion. When foods begin to enter the stomach, the contents of these glands, which are considerable, appear on the surface of the membrane, the muscular coat becomes active, and the food is manipulated in such a manner that all portions of it shall be acted upon by the gastric juice. The food is then changed in character and made to assume a uniform mass, part of which in a fluid state is absorbed by the capillaries of the stomach, the remainder passes down to be digested and absorbed in the small intestine. The position and shape of the stomach change considerably, particularly during digestion, and consequently can not be well mapped out externally. However, the upper or cardiac end is about on a level with the seventh rib on the left side.

The stomach is collapsed when empty, but when full it lies immediately against the abdominal walls, and also presses upward, and may interfere with the action of the heart and lungs. The distress that often occurs after a full meal is often due to distention of the stomach.

The gastric juice is an acid secretion, differing from all of the other digestive fluids, which are alkaline. There is a large quantity secreted in twenty-four hours, probably about fourteen pints.

The *small intestine* (Fig. 28), the most important organ of digestion, is the next portion of the alimentary tract, and begins at the pyloric extremity of the stomach. It is a tube about sixteen feet long and one inch in diameter, and composed of the three coats arranged in the same manner as the coats of the stomach. It is divided into three portions: the first and shortest portion being called the duodenum, which is about eight inches long; this is followed by the second portion or jejunum; the last portion being known as the ileum. The jejunum is generally empty after death, and receives its name from this fact.

In the mucous membrane of this portion of the canal

50 PROMPT AID TO THE INJURED.

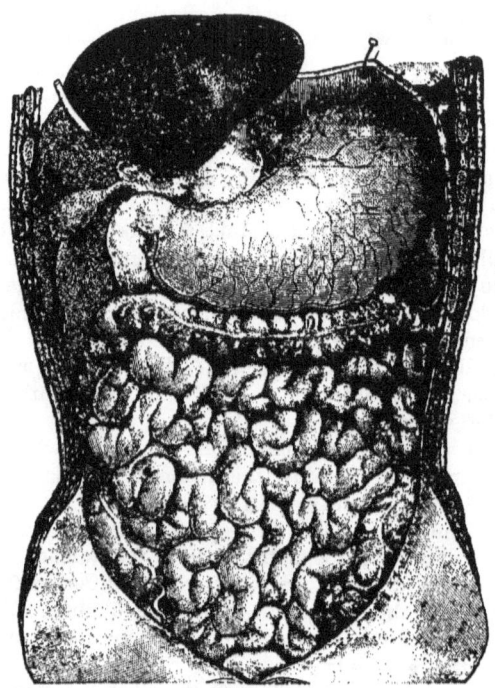

Fig. 28.—Position of abdominal contents.

are found glands yielding secretions which tend to dispose of certain articles of food incompletely digested, or unaffected by the secretion of the stomach.

The *large intestine* begins on the right side of the body, and is continuous with the ileum, or final portion of the small intestine, and, in its construction, is similar to that portion of the canal, although its diameter is considerably larger. It is divided into ascending, transverse, and descending colon, sigmoid flexture, and rectum, the latter terminating at the opening known as the anus. The

ALIMENTATION AND DIGESTION. 51

digestive power of the large intestine is very feeble. It is mainly a temporary receptacle for undigested and refuse matter which is to be discharged from the body. The transverse colon, which is about on a level with the umbilicus or "navel," is often the seat of intense pain known as "colic."

The *liver* (Fig. 29) is the largest organ in the body. It is situated in the abdominal cavity, below the diaphragm and above the stomach, principally on the right side, its long-

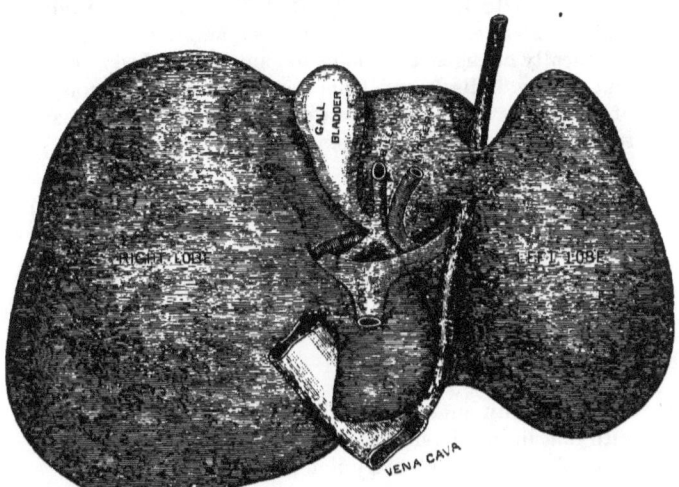

FIG. 29.—Under surface of the liver, showing the gall-bladder and a section of blood-vessels.

est diameter being from right to left. This organ, moderately filled with blood, weighs from four to four and one half pounds, being about twelve inches long, six inches wide, and three inches at its thickest portion. The surfaces of the liver are smooth and the edges are rounded, and it is of a dark-brown color. In the normal or healthy condition the liver extends downward on the right side as far as the lower border of the ribs. However, under certain conditions, as in tight lacing, it may be pressed below this point. Above and on the same side the liver extends to about one inch below the nipple. It also reaches to the left side of the median line of the body an inch and a half to two inches beyond the margin of the sternum and downward to midway between the lower end of the breastbone and the "navel." The liver has three distinct and separate functions: First, it renders very important aid to digestion, through the action of the bile which it secretes. It also produces sugar, which becomes chemically changed and helps to generate the body-heat; and, finally, it discharges from the body a certain amount of waste matter, or, in other words, it excretes as well as secretes. The glands of the skin also secrete as well as excrete. The bile, already referred to as the secretion of the liver, is a yellowish-brown alkaline fluid, and very bitter to the taste, and is discharged from the liver through a small tube, about the size of a goose-quill, which opens into the duodenum, or first portion of the small intestine; in this manner it reaches the food upon which it is to act. Connected with the under surface of the liver is a pouch or sac about four inches long and one inch in width, called the gall-bladder (Fig. 28). It is formed of fibrous tissue and lined with mucous membrane, which is continuous with that lining the bile-ducts. The gall-bladder acts as a reservoir where bile accumulates, being expelled during digestion, when a large amount is required. The secretion of bile is continuous, although it is only discharged into the intestine during digestion.

The *pancreas* or "belly sweet-bread"* (Fig. 30) is a

* There are three kinds of sweet-breads, viz.: the thyroid-gland, or throat sweet-bread, which is tough, almost like India-rubber; the pancreas, or belly sweet-bread, which is more tender, and is quite commonly used; and the thymus-gland, or breast sweet-bread, which exists only in young animals, wasting away as they grow up. The thymus-gland, situated just behind the upper portion of the breastbone, attains its greatest size in human beings at the age of two years, and disappears before the sixteenth year. Its use is not known. This

glandular organ situated at the upper and back part of the abdominal cavity, opposite the first lumbar vertebra, and

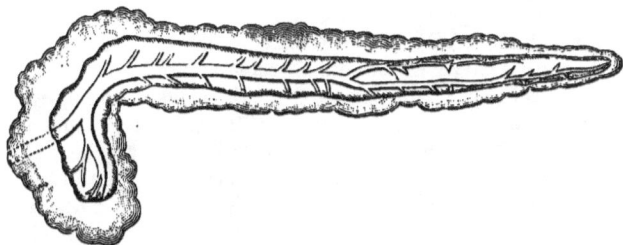

Fig. 30.—The pancreas, partly cut away, so as to show the duct which collects the pancreatic juice and empties it into the duodenum.

is mainly on the left side, and behind the stomach. The pancreas is about seven inches long and one inch in thickness, secretes an alkaline fluid, the pancreatic juice, and discharges it through a small duct which enters the duodenum at the same opening which receives the tube conducting the bile from the liver. One function of the pancreatic juice is to prepare the oil and fats of the food for absorption by the process known as emulsification. As the result of this transformation the oils and fats are converted into chyle, a fluid resembling milk in appearance, which is absorbed by lymphatic vessels along the intestinal tract, and named lacteals or milk-carriers, on account of their white appearance when filled with chyle. This fluid is carried by the lymphatic vessels to the receptaculum chyli, a pouch situated in front of the body of the second lumbar vertebra, and thence, by means of the thoracic duct, to a large vein on the left upper side of the chest, through which it reaches the circulation, and forms a very important element of nutrition. The pancreatic juice also converts meats, albumin, etc., into peptones, and changes starch into sugar.

gland, taken from calves and lambs, is the most tender and palatable sweet-bread of all. (Tracy.)

CHAPTER VI.

KIDNEYS, BLADDER, SKIN, SPLEEN.

KIDNEYS.

THE *kidneys* (Fig. 31) are excretory organs, and consist of two large tubular glands situated in the back part of the abdominal cavity in the lumbar region on each side of the spinal column. They extend from the eleventh rib downward nearly to the upper borders of the pelvis or "haunch-bones." The right kidney is a little lower than the left. They are bean-shaped, about four inches in length and two in width, and each weighing about four to six ounces. It will be remembered that the excretory organs simply abstract from the blood elements of waste tissue, and discharge the same from the body. They exert very little change in this material, but act mainly as an outlet for it. The material excreted by the kidneys is known as the urine, and consists of water holding in solution waste and poisonous matter, as urea, etc. The amount of urine removed by the kidneys in twenty-four hours is about three pints. It is an acid fluid when first discharged, but usually becomes alkaline as the result of decomposition. The ureters are two small tubes (one from each kidney) about the size of a goose-quill, and sixteen or eighteen inches long, which conduct the urine from the kidneys to the bladder, entering that organ at its lower portion. Oftentimes, small concretions, varying in size from grains of sand to those of larger diameter, called renal calculi, familiarly known as "gravel," are formed in the kidneys, and during the passage of these little bodies through the ureters on the way to the bladder, the larger ones may cause such irritation and pressure along these tubes as to produce the most intense pain. This condition has received the name of renal or kidney colic. Sudden and intense pain on the side, along the course of the ureter involved, would be the principal symptom. The sudden beginning of the pain would denote the entrance of one or more of the little bodies into the ureter, and the abrupt ending of the pain indicate that they had entered the bladder.

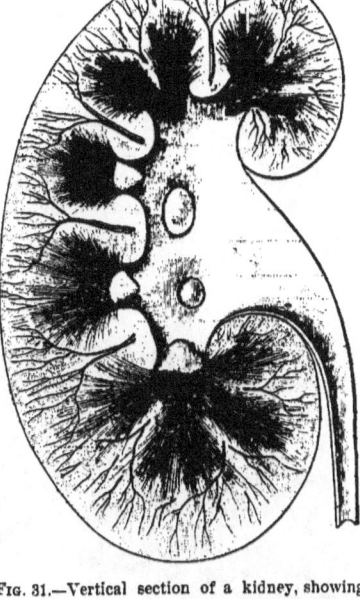

FIG. 31.—Vertical section of a kidney, showing the minute tubes which excrete the urine; also the pelvis, or basin of the kidney, which receives the urine before it passes downward to the bladder. The beginning of a ureter is also shown.

THE BLADDER.

The *bladder*, or reservoir which receives the urine, is a sac having three coats or layers arranged in a manner similar to that of the layers of the stomach and intestine—fibro-serous externally, muscular tissue forming the middle coat, and the internal being composed of mucous membrane. When moderately distended the bladder is about five by three inches and contains about one pint; however, it may hold considerably more. When a sufficient amount of urine has accumulated, it is discharged from the bladder, mainly by the contraction of its muscular walls; this act, which in the beginning is voluntary, usually occurs four or five times during the twenty-four hours. The bladder is situated in the lower part of the pelvic cavity, just behind the pubic bone.

SKIN (INTEGUMENT).

This structure, which covers all portions of the body and protects the deeper parts, is an organ of excretion and secretion, and also of the sense of touch, the latter function being highly developed at the finger-ends. The power of absorption through the unbroken skin of the human being is generally believed to be extremely slight. The skin is composed of two layers, the outer and inner; the former being called the epidermis, cuticle, scarf, or false skin, and the latter receiving the name of derma, corium, or true skin (Fig. 32). The epidermis contains neither blood-vessels nor nerves, and is simply a scaly layer which protects the true skin underneath, and becomes easi-

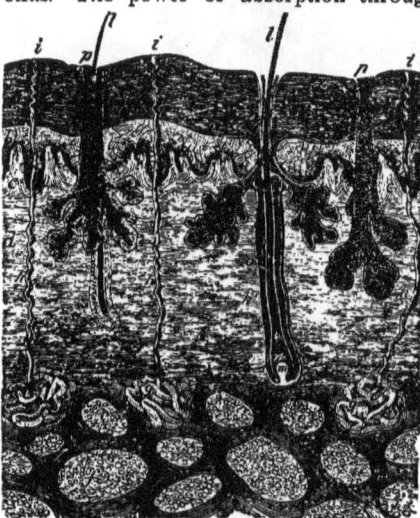

FIG. 32.—Vertical section of the skin, magnified. *a*, Scarf-skin; *b*, pigment-cells; *c*, papillæ; *d*, true skin; *e, f*, fat-cells; *g*, sweat-glands; *h*, outlets of sweat-glands; *i*, their openings on the surface of the skin; *k*, hair-follicle; *l*, hairs projecting from the skin; *m*, hair-papilla; *n*, hair-bulb; *o*, root of hair; *p*, openings of oil-glands.

ly detached from it under certain conditions. A blister shows the separation of the two layers of skin, with a small amount of fluid between them. The derma or true skin is richly supplied with blood-vessels, nerves, and lymphatics, and in this layer reside the functions above alluded to. The skin contains millions of minute tubes known as sudoriferous or "sweat" glands; also the sebaceous or fat glands, and hair-follicles. The sweat-glands, by constantly removing from the body a watery vapor known as perspiration or "sweat," in which they are aided by evaporation, help to regulate the body temperature. It is the wonderful activity of these organs, and the large quantity of perspiration thrown off, that enables one to endure a very high degree of dry heat. In addition to the action just described, they have the important function of eliminating from the system waste material similar to that excreted by the kidneys (urea); for this reason the skin, through the sweat-glands, is regarded somewhat as a supplementary organ to the kidneys. In warm weather, when the skin is most active, the amount of perspiration is largely increased and the amount of urine correspondingly diminished, while in winter the condition is reversed. An average amount of perspiration formed in twenty-four hours is about two pints. It is very important that the similarity in action of the skin and kidneys should be recognized, and that the one may be induced to relieve the other, for, when the kidneys are impaired, either temporarily or permanently, much relief may be obtained by stimulating the skin.

The importance of keeping the skin in a healthy condition by proper clothing, exercise, and bathing, can not be overestimated. The product of the fat or sebaceous glands protects the skin, and keeps it more or less oily and pliable.

With few exceptions the hair-follicles are found in all portions of the body; in some parts, however, the hair is so fine as to be hardly noticeable. The hairs, and also the nails, are regarded as appendages of the skin, affording additional protection.

SPLEEN.

The spleen is a soft and spongy organ, situated in the abdominal cavity, on the left side, near the cardiac end of the stomach, and extends from the ninth to the eleventh ribs. It measures about five inches long by three inches wide, and two inches thick, weighing about eight ounces. It is called a ductless gland, not having the characteristics of a secretory or an excretory organ. Its function has not yet been definitely settled

CHAPTER VII.

NERVOUS SYSTEM.

THE activity of the mind and body, the correct working of the several organs, and the sympathy existing between the different parts of the organism, depend upon the nervous system, which is divided into the cerebro-spinal axis and the sympathetic system.

CEREBRO-SPINAL SYSTEM.—This system comprises the brain, spinal cord, and nerves (Fig. 34).

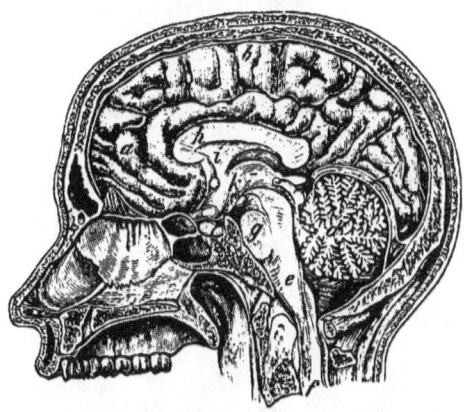

FIG. 33.—The brain inclosed in its membranes, and the skull : *a, b, c,* convolutions of the cerebrum; *d,* the cerebellum ; *e,* medulla oblongata ; *f,* upper end of the spinal cord ; *g,* pons Varolii ; *h, i, k,* central parts.

The *brain* (Fig. 35), which is the seat of the intellect, the will, and the emotions, is contained in the cranial cav-

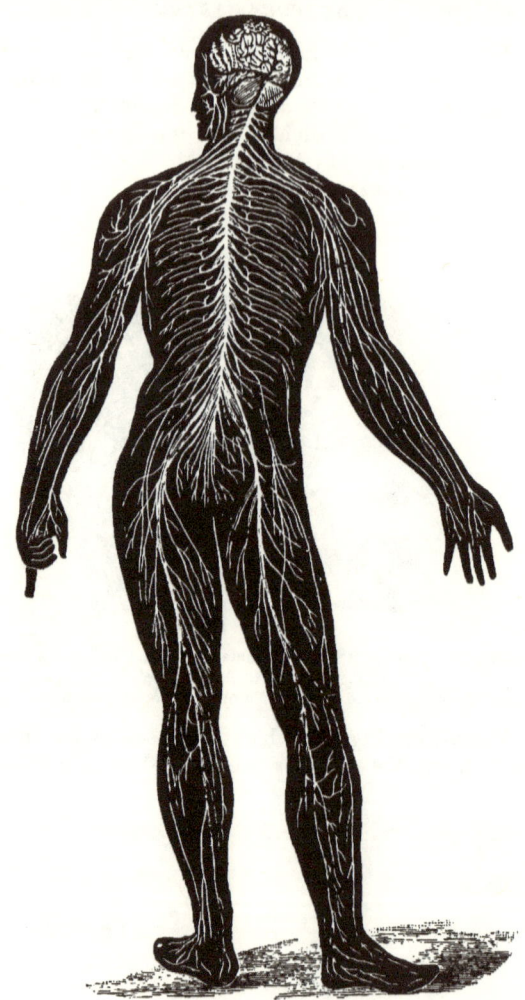

Fig. 34.—The cerebro-spinal system of nerves.

ity (Fig. 33). It is divided ino four principal parts, viz., the cerebrum, the cerebellum, the pons Varolii, and the medulla oblongata.

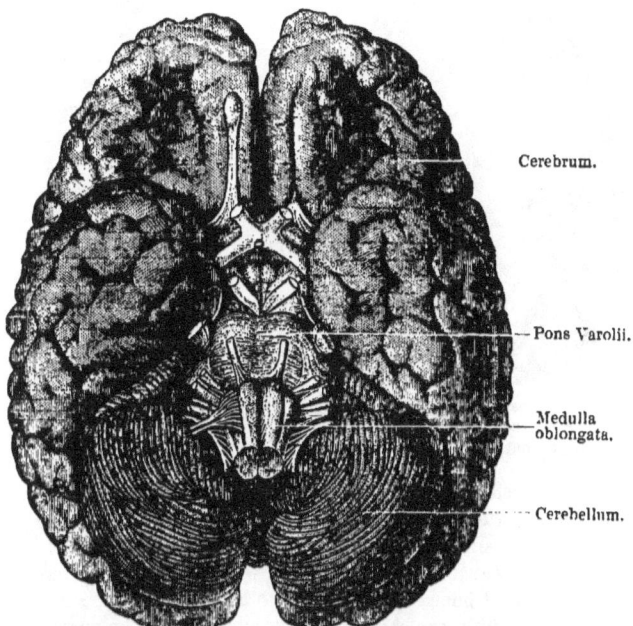

FIG. 35.—Under surface of brain, showing the cerebrum, cerebellum, pons Varolii, and medulla oblongata.

The cerebrum (Fig. 36) constitutes over four fifths of the entire brain. It is ovoid in form, being divided into two portions by a deep groove, running from before backward, and occupies the upper and greater portion of the cranial cavity. It is composed of gray and white matter, about the consistence of "sweet-breads"; the white matter forms the internal and greater portion of the cerebrum, the gray matter forming the external layer. The surface of the cerebrum is not smooth, but thrown into folds, called convolutions. This arrangement increases its area, and consequently its function. The convolutions are superficial and not so well marked in the brain of an infant, but grow deeper and larger until the brain has reached its full development, at about the fortieth year. The cerebrum is the seat of the mind and of its different functions which distinguish man from the lower animals. The exact point in the cerebrum where these functions are located may at some future time be definitely settled, considerable progress having already been made in this direction.

FIG. 36.—Upper surface of the cerebrum, showing the convolutions of the brain and its double structure.

The cerebellum, or little brain, corresponds in structure quite closely to the cerebrum, with which it is connected, being situated beneath it and at the back part of the cranial cavity. The cerebellum regulates and keeps in perfect harmony the different movements of the body, particularly of the extremities (Figs. 34 and 35).

The pons Varolii, or "bridge," is a small portion of the brain, and situated in front of the cerebellum. It binds together the different parts of the brain already enumerated, and also transmits the different nerves passing between the brain and spinal cord (Fig. 35).

The medulla oblongata is below the pons Varolii, and appears to be the upper end of the spinal cord somewhat expanded. In the substance of this organ about three fourths of the motor-nerve fibers passing from the brain to the spinal cord cross each other or decussate, decussation of the remainder taking place in the spinal cord; consequently a motor-nerve fiber having its origin in the right side of the brain crosses to the left when it reaches the medulla oblongata, and becomes identified with the left side of the spinal cord, and furnishes motion to that side of the body. This will explain why an injury to one side of the brain will produce paralysis on the opposite side of the body, as in apoplexy. The medulla also presides over the function of respiration.

The human brain weighs about fifty ounces, being heavier than that of any of the lower animals, with the exception of the elephant and the whale. The brain of the elephant weighs about eight pounds, that of the whale somewhat less. The human brain probably reaches its highest development at about the fortieth year. After this period it gradually diminishes in weight, about one ounce in ten years, so that in old age both the size and function of the brain are considerably lessened.

The deep convolutions spoken of are characteristic of the human brain, being present in a lesser degree in the lower animals.

The *spinal cord* is that portion of the cerebro-spinal axis contained in the spinal or vertebral canal. It is cylindrical in shape, extending downward from the medulla oblongata to the first lumbar vertebra, being from fifteen to eighteen inches long, and is composed of gray and white matter, the white matter, however, being outside and consisting of nerve-fibers which act as conductors of sensory and motor influence. The motor fibers of the cord are found in the front and at the sides, and the sensory fibers in the back or posterior portion.

The spinal cord, through the medium of its branches, to be presently described, transmits nerve influence to and from the brain. It is, to a certain extent, capable of acting as a separate nerve-center and generating force independently of the brain.

Nerves are classified as those having a motor and sensory influence, and nerves of special sense. Motor nerves are fibers which conduct from the brain (and to a certain extent, from the spinal cord) the necessary force to animate muscular fibers, thus producing the different movements of the body; motor nerves therefore transmit an influence from within outward. Sensory nerves are fibers which convey from different portions of the body certain sensations to the brain. For instance, when a finger is burned or injured, the sensation of pain experienced indicates an impression made upon a sensory nerve at the point of injury, and which is received at the great nerve-center or brain. Nerves of special sense, as the name implies, have separate functions. They do not transmit motion or common sensation just described, but preside over the special senses, as sight, hearing, taste, etc.

Cranial Nerves.—From the under surface or base of the brain, on each side, are given off twelve nerves, which are known as the twelve pairs of cranial nerves. They are composed of white, glistening fibers, and are numbered anatomically from before backward. These nerves are so very important that a brief description of them will be necessary.

The *first*, or olfactory nerves, supply the special sense of smell, and are distributed to the mucous membrane lining the nose. The *second*, or optic nerves, supply the special sense of sight, and are distributed to the eyes. The *third, fourth,* and *sixth* are motor nerves, and animate the muscles moving the eyeballs. The *fifth* are the largest of the cranial nerves, and have a mixed function; the sensitive branches supply common sensation to the teeth, nearly all of the face, the mucous membrane of the eye, nose, and portions of the mouth and throat; the motor branches supply the muscles of mastication. The *seventh* supply, with the exception of those directly concerned in mastication, the muscles of the face with motion. A small branch supplies the special sense of taste to the anterior portion of the tongue. The *eighth*, or auditory nerves, furnish the sense of hearing. The *ninth* supply the posterior part of the tongue with the special sense of taste, and also common sensation to the tongue and pharynx. The *tenth*, which at its origin is a sensory nerve, receives motor branches from the seventh, eleventh, and twelfth, and first and second cervical nerves. The motor, sensory, and mixed branches of this nerve go to the pharynx, larynx, œsophagus, heart, stomach, small intestine, liver, spleen, and kidneys.

The *eleventh* furnish part of the motor supply to one muscle of the neck in front and to a muscle of the shoulder and neck posteriorly (each side). By the branches which join the tenth, these nerves supply the muscles of the larynx concerned in the production of the voice; they also contribute to the regulation of the action of the heart.

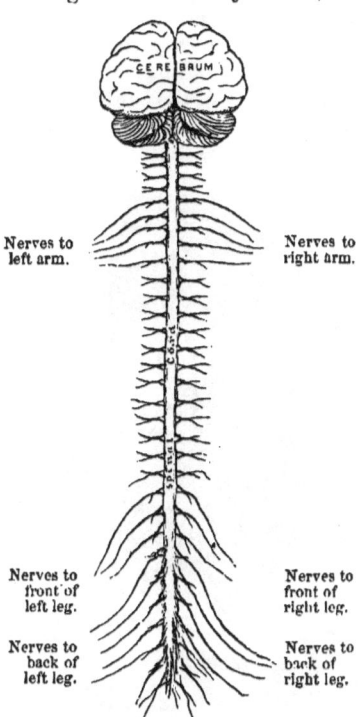

Fig. 37.—Brain and spinal cord, with the thirty-one pairs of spinal nerves.

The *twelfth* are motor nerves, and preside over the movements of the tongue.

The cranial nerves that supply structures outside of the cranial cavity pass through small openings in the skull on the way to their destination.

Spinal Nerves (Fig. 37).—There are thirty-one nerves given off from each side of the spinal cord, each nerve forming with its fellow on the opposite side a pair similar to the arrangement of the cranial nerves. They also resemble those nerves in appearance. Each pair of nerves are composed of fibers arising from the anterior and posterior columns of the cord, and consequently contain motor and sensory fibers. At a short distance from the spinal cord the nerves divide into anterior and posterior branches, the anterior branches supplying the trunk and extremities, diaphragm, and certain organs. The posterior or smaller branches supply principally the muscles and skin of the back.

The brain and spinal cord are each protected by a closed sac, surrounding them, after the manner of the pericardium which envelops the heart, and the pleura covering the lungs.

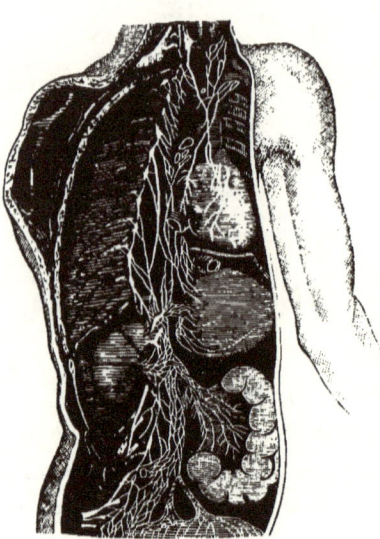

FIG. 38.—The sympathetic or ganglionic nervous system.

The portion of the nervous system just described presides over the functions of animal life—the intellect, general sensation, the special senses, and motion.

SYMPATHETIC SYSTEM.— Organic life, or that which relates to secretion and excretion, the proper distribution of blood and nutrition to the tissues, the control of the involuntary muscular tissue which is found in the alimentary tract, blood-vessels, etc., must be unaffected by the outside world. Were it subject to the will, and to such influences as govern the cerebro-spinal axis, the functions of organic life would soon be disturbed, and the most serious consequences would ensue. This system must work on uninterruptedly, in health and disease, night and day— at all times. For this purpose Nature has supplied a separate and independent nerve-force known as the *sympathetic* or ganglionic system (Fig. 38), which consists of a series of little bodies or ganglia which begin at the under surface of the brain, and continue downward on each side of the spinal or vertebral column to its lower extremity. Although this system is independent of the cerebro-spinal axis, in its special function, it communicates with, and accompanies the cranial and spinal nerves to organs which are supplied by both systems. It also brings the different organs in communication with each other, and admits of the sympathy which exists between them. This relation between the different organs explains why a disordered stomach will produce headache, or why death may follow a blow at the pit of the stomach, which injures the large sympathetic ganglia back of this organ, the shock conveyed to the heart being so great as to arrest its action.

CHAPTER VIII.

BANDAGES, DRESSINGS, ETC.

BANDAGES are used to retain dressings in position, arrest hæmorrhage, and support and render immovable different portions of the body. For general use they are divided into roller, and Esmarch or triangular bandages.

The roller bandage when properly applied is not only a valuable means of employing this form of dressing, but is also attractive in appearance; considerable skill, however, is essential in order to make its application effective. The choice and preparation of material necessary for its construction render its use impossible at all times and places; for this reason the Esmarch or triangular bandage, which can be made from a handkerchief, shirt, or some other material that can be procured at once, is accepted as being best adapted for emergencies.

The materials utilized for bandages include linen, flannel, calico, and muslin. Rubber tissue is also often used by surgeons for special purposes, as support in varicose veins of the leg, and flannel where warmth is required. For ordinary purposes, unbleached muslin of a medium texture is the best fabric that can be employed.

Bandages should not contain starch, nor should they be pieced, as either condition produces considerable irritation of the skin, and also prevents their proper application.

ROLLER BANDAGES are usually made from three to eight yards long and from one to six inches wide, depending upon the part of the body to which they are to be applied—one inch for the fingers, three inches for the upper extremity, four inches for the lower extremity, and five or six inches for the chest and abdomen. Bandages should always be *torn* from the piece, unless the material from which they are constructed is very thin. The selvedge along the edge of the fabric should always be removed before the bandages are rolled.

In rolling a bandage the following directions are to be observed: One end of the strip should be folded for a distance of about six inches; this lap should then be folded on itself, and the process continued until the folded portion assumes the form of a roll or cylinder, which should then be held by the thumb and index-finger of the right hand, the loose strip of the bandage lying over the back of the left hand between the thumb and index-finger. The hand holding the bandage should now be turned toward the right, when the roll will begin to increase in size; this manipulation is to be continued until the bandage is entirely rolled; it is very important that the bandage should be wound tightly, particularly at the beginning, otherwise it can not be evenly adjusted (Fig. 39).

FIG. 39.—Method of rolling a bandage.

A bandage usually has one head or roll, although it may be made double-headed (by rolling from both ends) for special purposes, as in the knotted bandage.

The application of a bandage should always be commenced by laying the *outer* surface against the skin. If used to retain a dressing, it may be begun at any part of an extremity, the soft dressing underneath preventing any undue interference with the circulation; but when the bandage is applied for support or pressure, it must be commenced at the extreme end of the limb and bandaged *toward* the body, otherwise the constriction, particularly if the bandage is drawn tightly, may be followed by strangulation and gangrene of the tissues below the bandage. A bandage should be closely applied to a limb, but not made tight, and the degree of pressure should be uniform. Should any evidence of strangulation occur, which would be manifested by swelling and discoloration, and also a reduced temperature of the limb below the bandage, the dressing must be at once removed. It is partly for this reason that the toes and the ends of the fingers are left uncovered in bandaging an extremity—that they may serve as an index to the general circulation of the arm or the leg. A bandage applied dry and wetted afterward is followed by considerable shrinking, and sometimes strangulation of the tissues.

A bandage should not be applied to a limb until the latter is placèd in the position in which it is to remain. For example, an arm should never be bandaged while straight, and then bent afterward, otherwise serious constriction will be made at the flexure.

For ordinary use the application of the roller bandage is divided into three methods: the *circular*, the *spiral reverse*, and the *figure of 8* (the latter including spica bandages).

The circular variety is the simplest manner in which a roller bandage can be applied, and, when possible, should always be employed. It is indicated when the diameter of the part is nearly uniform, and consists of a succession of circular turns from below upward, each turn overlapping the upper third of the one below.

The spiral reverse bandage is used where the diameter of the part becomes increased or decreased in size, as in the forearm (Fig. 40).

BANDAGES, DRESSINGS, ETC.

The figure-of-8 bandage is used about joints or where an abrupt enlargement occurs. The hip and shoulder spica are varieties of this bandage. The manner in which the different layers cross each other give to it something of the appearance of a figure of 8; hence the name.

In order to illustrate the different varieties of bandages named above, it may be supposed, for example, that the left lower extremity, including the hip, is to be bandaged, which generally necessitates the use of the different forms already enumerated.

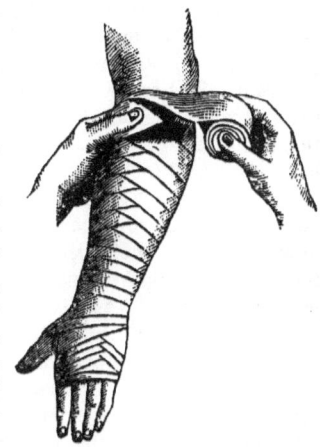

FIG. 40.—Spiral reverse bandage.

Several bandages, having a width of about three and a half to four inches, are procured. The operator places himself directly in front of the limb to be bandaged, which is somewhat elevated, the patient being seated or in bed. The operator, holding the bandage in his right hand, unrolls about six inches of it, the outer side of which is laid obliquely across the dorsum or top of the foot from within outward, and continued around the lower part of the ankle to the inner side, and then again over the dorsum of the foot, crossing the first portion applied, and carried downward to the base of the toes on the outer side; from this point two or three circular turns are made around the foot, extending upward, and thence again to the ankle, where the circular form is employed until the calf of the leg is reached. If the circular turns are continued at this point, the lower border will be seen to gape ; for this reason the spiral reverse method should now be substituted (Fig. 40). From the last circular turn, the bandage should be carried spirally upward and outward for about five or six inches. The operator then presses with the thumb of his left hand the body of the bandage against the median line of the leg, to prevent it from slipping, while the right (holding the bandage) is brought toward the operator with a slight inward turn, thus folding downward (reversing) the upper edge of the bandage to a point just above the thumb of the left hand, and held for a moment in this manner until the disengaged fingers of the left hand are carried behind the leg and receive the roller or cylinder from the right, which, now being free, retains the fold or reverse in position until the bandage is again passed to it for another reverse. This is to be continued (each turn overlapping the upper third of the turn below) until the portion of the leg above the calf is reached, where the circular turns can be made up to the lower edge of the patella or "knee-cap," where the figure-of-8 bandage is begun in the following manner: after a circular turn, the bandage is carried obliquely over the patella from within outward and upward, making a circular turn above the knee, then to the inner side of the leg and downward and outward, crossing the portion of the bandage over the patella. A circular turn is now made around the knee below the patella (to retain the spiral or oblique turns already made), and then carried again obliquely upward and outward, overlapping the upper half or one third of the preceding turn in this direction, and continued in this manner until the entire knee is covered. Above the knee, the circular form can be used for a short distance and then changed to a reverse as it ascends the thigh, and continued to a point just below the groin; from this point the application of the hip spica begins, the bandage is carried upward and around the hip to the opposite side of the body, just above the great trochanter of the right femur; this bony prominence marks the upper extremity of the thigh-bone, and can readily be found, particularly if the patient be made to move the leg.

Following this course, the bandage is carried over the belly to the point where it started from (left thigh), crossing the first turn and passing downward and backward around the thigh. This procedure is to be repeated in an ascending manner until the hip is covered. The bandage, when applied, should not extend above the upper border of the haunch-bones (crest of the ilium). It will be noticed that, in bandaging the foot, the toes and heel are left uncovered. This is the usual mode of applying the bandage, unless an injury exists at these parts.

The shoulder spica is applied in a manner similar to the spica of the hip. When the bandage is carried to the opposite side, it passes beneath the arm and back again over the shoulder, from where it began. It also is an ascending bandage (Fig. 41). In the spica bandage of the hip and shoulder, the overlapping is greater at the opposite side than at the hip or shoulder which is being covered. The spica bandages (hip and shoulder) are used to retain dressings and apparatus.

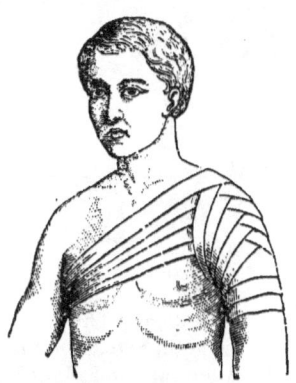

Fig. 41.—Ascending spica of the shoulder.

The double-headed knotted bandage is made of a strip of muslin about eight yards long and two inches wide, and rolled into two heads or cylinders. The knotted bandage is generally used to arrest hæmorrhage from the scalp by making pressure upon the temporal artery, which is situated just in front of the ear and extending upward into the scalp. The pulsation of this vessel can easily be felt. Before applying the bandage, a firm compress should be placed over the artery at a level with the upper border of the ear. A piece of cork, for example, about half an inch thick and the diameter of a silver half-dollar, should be enveloped in a piece of soft muslin and applied over the artery. The operator then, holding a roller in each hand, places the outside of the bandage against the compress and carries one roller around the head just above the eyes in front, and the other below the occipital protuberance or "bump" on the back of the head, and to the opposite temple ; at this point the hands of the operator change rollers and return them to the compress, over which they form a knot by twisting and changing their direction, one roller being carried over the top of the head and the other beneath the chin, and meeting at the opposite side. At this point the rollers are again changed and returned to the compress, over which a second knot or twist is made in the same manner ; two or three knots over the compress are usually sufficient to arrest the hæmorrhage (Fig. 42).

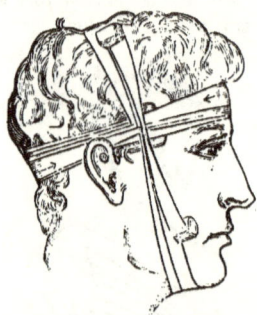

Fig. 42.—Knotted bandage.

There are a number of bandages which can not be classed with either the roller or triangular bandage. . The following are examples :

The FOUR-TAILED BANDAGE is made of a strip of muslin about one yard and a half long and four inches wide, folded and torn from the ends to within two inches of the center of the bandage. One pair of tails is usually made broader than the other. One purpose of this bandage is to support the lower jaw after a fracture or dislocation. It is to be applied by placing the center of the bandage against the chin, with the wide tails below, when the latter are turned upward and tied on top of the head; the upper or narrow tails are carried backward and tied at the nape of the neck (Fig. 43) ; two handkerchiefs, each folded in the form of a cravat, may be used for this purpose. A four-tailed bandage, for the purpose of protection and retaining dressings about the head, can be made of a piece of muslin, about one yard and a half long and one foot or more in width, folded, and torn from the ends to within six inches of the center. The bandage is placed on the head, the anterior tails being carried backward and tied at the nape of the neck, while the posterior tails are tied under the chin (Fig. 44). If it is required that the bandage should be applied to the back of the head, the anterior or upper tails are tied beneath the chin, while the posterior or lower tails are tied upon the forehead.

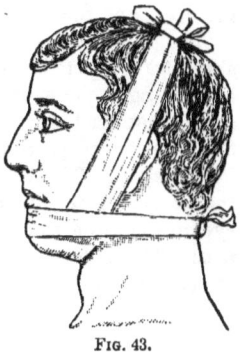

FIG. 43.

A four-tailed bandage may also be used at the knee. It should be made of a strip of muslin about eight inches wide and one yard long, and applied by placing the center of the bandage over the patella; the upper tails are carried backward, crossed behind the knee, then brought forward and tied in front below the patella; the lower tails are manipulated in a similar manner, but tied above the patella.

FIG. 44.

The entire head may be covered by the application of a six-tailed bandage, made from a piece of muslin about forty inches long and fifteen inches wide and folded. The ends should be torn into three tails, extending to within six inches of the center of the bandage; the middle tail should be made wider than the outer ones. The bandage being placed on the head, the middle tails are tied under the chin, the anterior at the nape of the neck, and the posterior on the forehead.

The large square handkerchief-bandage is a very valuable means of covering the head and neck, and is applied by folding a piece of muslin about one yard square, in such a manner that the upper portion or layer is narrower than the lower portion, their borders being two or three inches apart. The bandage is so placed on the head that the border of the upper portion is on a level with the eyebrows, while the lower layer rests on the tip of the nose. The ends of the upper portion are tied beneath the chin; the border of the lower portion is then turned up-

FIG. 45.

ward against the forehead, thus exposing the eyes, while the ends of this portion are carried backward and tied behind the neck (Figs. 45 and 46).

ESMARCH OR TRIANGULAR BANDAGES (see diagram, Fig.

47.)—The triangular bandages, which have been brought to the notice both of the medical profession and the laity by Prof. Esmarch, of Germany, are not only easily made from material which is generally procured without difficulty, but are effective as a dressing and easily applied, and act as a substitute for the roller bandage in all cases except where uniform pressure is the essential element.

If a choice of material is permitted by the existing circumstances, the unbleached muslin used for roller bandages should be selected; however, a large handkerchief, or a shirt or skirt, may be used.

FIG. 46.

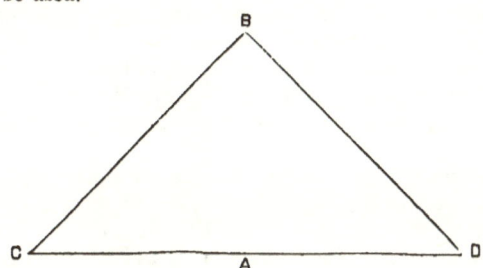

FIG. 47.—Esmarch bandage. A, base; B, apex; C, D, basal ends.

It will be unnecessary to enumerate the sizes given for the different triangular bandages. A proper knowledge of their application will sufficiently indicate this when being used on different portions of the body.

The ends of a triangular bandage should be fastened either by tying with a "reef" or flat knot, or with safety-pins; the knot, however, is generally used (see KNOTS).

Head (Figs. 48, 49, and 52).—The base (*A*) is placed downward over the brow, with the apex (*B*) at the nape

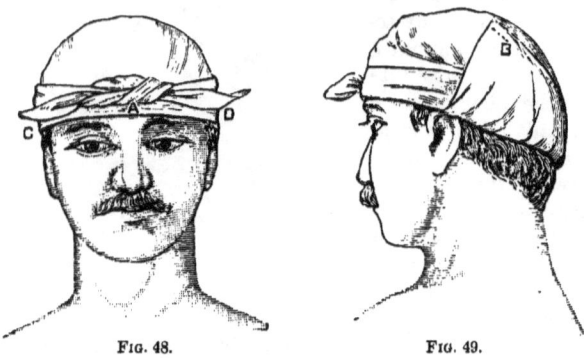

FIG. 48.   FIG. 49.

of the neck. The basal ends (*C* and *D*) are carried backward above the ears and crossed over the apex *below* the external occipital protuberance, the "bump" on the back of the head; this prevents the bandage from slipping upward; the basal ends are then returned to the front and tied, the apex being turned upward and pinned to the body of the bandage.

Shoulder (Fig. 50).—The triangle should be applied to the shoulder by placing the base (*A*) downward across the middle of the arm, the apex (*B*) being turned upward against the neck. The basal ends (*C* and *D*) are carried to the inner side of the arm, crossed, and returned to the outside and fastened. The apex (*B*) is tied or pinned to a cravat or sling placed around the neck.

Hand (Fig. 50).—In injury to the posterior part, or

dorsum, the hand may be bandaged by placing the base of the triangle (*A*) upward at the back part of the wrist, the hand lying on the bandage; the apex (*B*) is turned over the fingers upon the palm and carried to the wrist; the basal ends (*C* and *D*) are then carried to the front and crossed, and returned to the back of the wrist and tied, or crossed again, and tied in front. In injury to the palmar

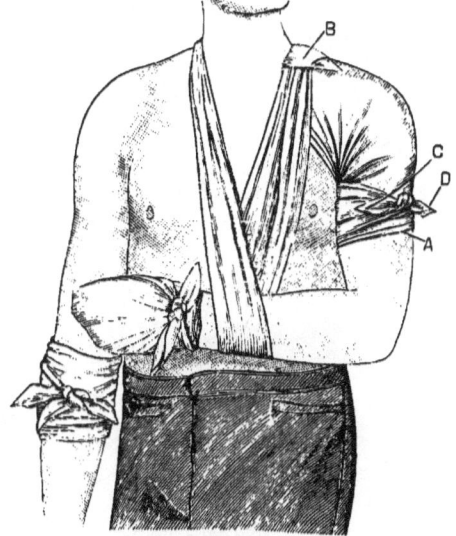

FIG. 50.

surface of the hand, the base should be first applied to the front of the wrist.

Chest (Figs. 51 and 52).—The triangular bandage is applied by placing the base (*A*) downward across the lower border of the chest, with the apex (*B*) over the shoulder of the affected side, the basal ends (*C* and *D*) being carried around the sides to the back, and tied together in such a manner that one end of the knot is longer than the other.

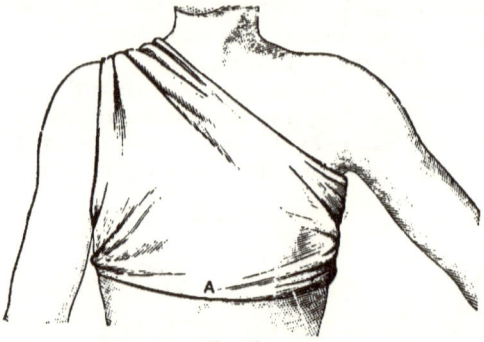

Fig. 51.

The long end is then tied to the apex, which has been carried over the shoulder.

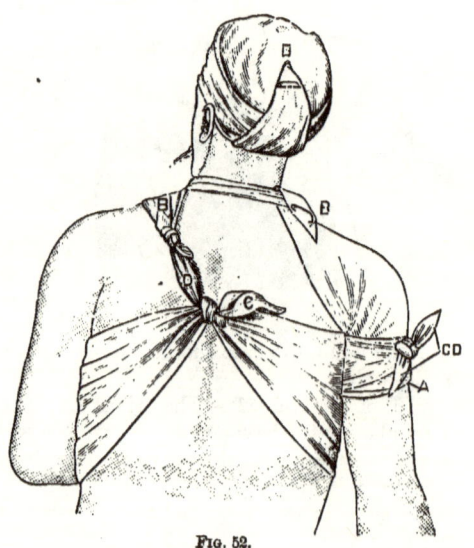

Fig. 52.

SLINGS. — The ordinary form of an arm-sling made of a triangular bandage is as follows (Fig. 53): For example, suppose that the left arm alone is injured; the base (A) is placed vertically along the outer border of the right side of the chest, with the basal end (C) thrown over the

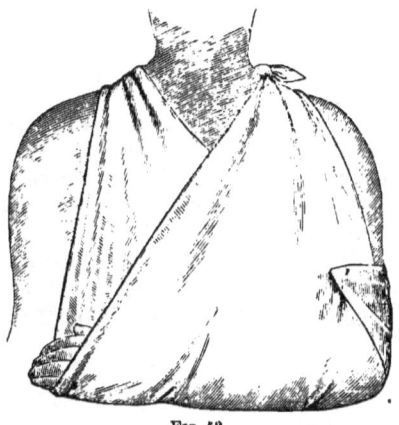

FIG. 53.

right shoulder, and the apex (B) placed behind the left elbow, the left arm being bent at a right angle and held in front of the bandage. The basal end (D) is now carried around the forearm and over the left shoulder, and tied to the basal end (C) at the back of the neck. The apex (B) is then carried around the elbow and pinned to the bandage in front.

Should it happen that the left shoulder or clavicular (collar-bone) region is injured as well as the arm, the sling should be

FIG. 54.

so arranged that the affected part is free from pressure, which may be accomplished in the following way (Fig. 54): The bandage is to be applied in a manner similar to the one just described, with this exception—instead of the basal end (*D*) being carried over the left shoulder, it is carried under the left arm, then upward across the back, and tied to the basal end (*C*) over the right shoulder.

If the right shoulder or clavicular region be injured as well as the left arm, the latter can be placed in a sling, leaving the affected shoulder (the right) uncovered by laying the basal end (*C*) over the left shoulder, the base (*A*) being carried from above obliquely downward and to the right, the apex of the bandage and the arm being in the same position as in the slings just enumerated. The basal end (*D*) is now carried under the left arm and upward over the back, and tied to the basal end (*C*) at the left shoulder (Fig. 55).

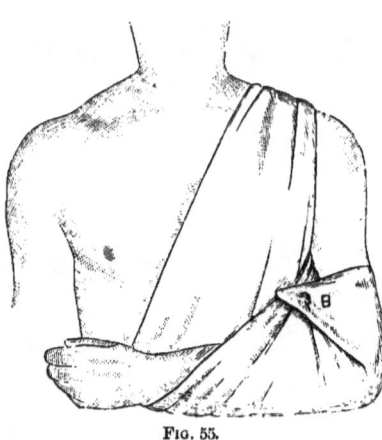

Fig. 55.

Hip.—The triangular bandage at this portion of the body is applied in very much the same way as at the shoulder. The center of the base (*A*) is placed downward across the middle of the thigh, the apex (*B*) being carried upward above the crest or upper border of the pelvis or haunch-bone, the basal ends (*C* and *D*) are carried around the thigh, and fastened at the outside. The apex (*B*) is attached above to a cravat around the waist (Fig. 56).

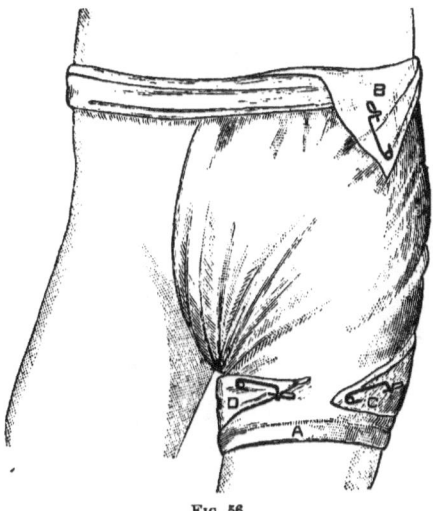

Fig. 56.

Foot.—The foot should be placed on the triangle with the base (*A*) backward, and laid behind the ankle, the apex being carried upward over the dorsum or top of the foot. The basal ends (*C* and *D*) are brought forward, crossed, then carried around the foot, and tied on top (Fig. 57).

CRAVAT-BANDAGES (Figs. 50, 58, 59).—A triangular bandage folded in the form of a cravat makes a very effective means of arresting hæmorrhage and retaining splints, dressings, and poultices. The width of the cravat, or the number of folds, depends upon the use for which

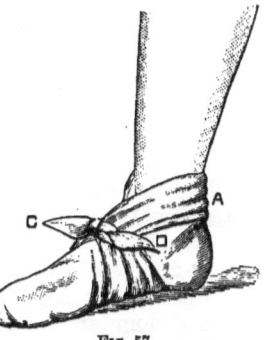

Fig. 57.

it is applied. The center of the cravat should be laid against the affected part, or on the poultice, or whatever the cravat is used to retain, the ends of the cravat being carried around

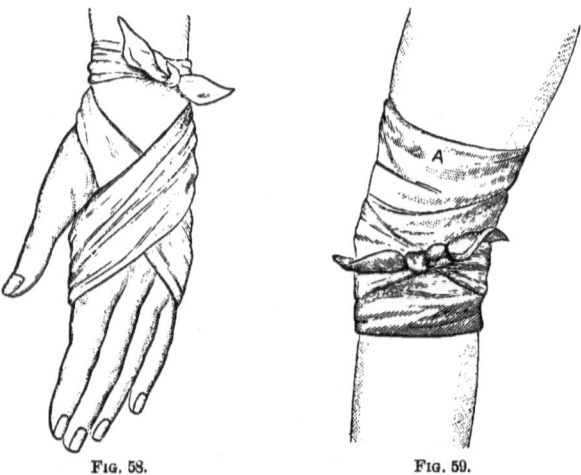

Fig. 58.   Fig. 59.

the limb and tied over the center of the base. When used to retain splints, they should be tied on the outer side of the limb and against the splint, thus preventing the knot from irritating the skin. Cravats are used to retain dressings about the head, eyes, ears, neck, etc. When used to retain a dressing in the arm-pit, the center of the cravat should be placed under the arm, and the ends carried upward and crossed over the shoulder, and tied in the axillary space of the opposite side, thus forming a figure of 8. In retaining dressing about the groin and in that vicinity, the middle of the bandage is placed at the inner and upper part of the thigh (the "crotch"), the extremities being carried upward and outward and crossed at the hip, the ends being brought over to the hip at the opposite side and tied.

The cravat may also be used as a sling for the arm where simply support, but not protection, is necessary.

KNOTS.—The ends of a triangular and cravat-bandage are fastened by tying with a "reef" or square or flat knot (Figs. 60 and 61). This

Fig. 60.—Reef knot.

Fig. 61.—Reef knot.

knot is very secure, and does not slip; it is used by surgeons in ligating vessels. However, unless care is observed, a "granny" knot (Figs. 62 and 63), which is quite insecure, is often substituted. The reef knot is made by holding an extremity of the bandage in each hand, and then passing the end in the *right* hand over the one in the left and tying; the end now in the *left* hand is passed over the one in the

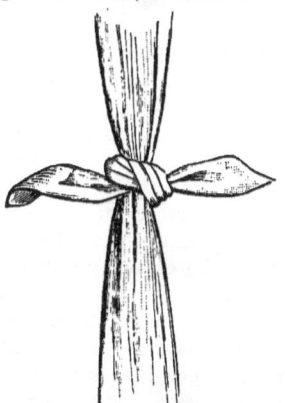

Fig. 62.—Granny knot.

Fig. 63.—Granny knot.

right and again tied. In the "granny" knot the end in the right hand is carried over that in the left both times and tied (Fig. 63).

FIG. 64.—Surgeon's knot.

A surgeon's knot is used in ligating blood-vessels, and is made by turning the ends of a ligature twice around each other before tying (Fig. 64), and then securing it as in the reef knot.

COMPRESSES.

Compresses or pads, as generally used, are of two kinds —simple and graduated.

A simple compress consists of a number of even folds of muslin or other similar material, applied to a part to assist in arresting hæmorrhage (by pressure), to hold the edges of a wound together, and also for protection. A graduated compress is made by diminishing the width of each successive layer until the compress has been made pyramidal or pointed, and having a base and an apex. When applied, the graduated compress is inverted, the apex being placed first in the wound. This form of compress is used to arrest hæmorrhage from deep wounds and cavities.

Compresses are made of antiseptic or sterilized gauze, lint, muslin, linen, flannel, etc., the antiseptic or sterilized gauze being the most desirable. If this can not be procured, the material selected should be clean, and, if possible, made aseptic by boiling or baking in an oven for fifteen or twenty minutes. Lint is objected to, as the surfaces—particularly one of them—are downy, and adhere to the wound. The common picked lint is objected to for the same reason. A compress may be prevented from adhering to the wound by smearing it with vaseline or oxide-of-zinc ointment. A compress made of a fabric such as gauze, which admits of absorption of the fluids from a wound, is more valuable than one which resists it.

A compress should not be allowed to remain applied after it has become offensive or soaked with discharges.

TAMPON.

A tampon or plug is made of the same material used for compresses, and, in addition, absorbent cotton, lamp-wicking, common muslin bandages, etc., are often utilized. A tampon has no special size or shape, but is formed in such a manner that it can be pressed deeply into a wound in order to arrest hæmorrhage, for which it is principally employed. It is of the utmost importance that a tampon should be clean, and also made antiseptic if possible, in order to prevent the entrance of poisonous germs. It should be remembered that, although it is necessary in order to arrest hæmorrhage by tamponing that the wound should be thoroughly filled from the bottom, too forcible manipulation may be followed by serious consequences, and that even the proper application of a plug usually prevents union by first intention. A tampon as well as a compress is kept firmly in place by a bandage, which exerts more or less pressure as the case requires. Without the indorsement of a surgeon, a tampon should not be left in a wound longer than twelve hours.

POULTICES.

Poultices are used to supply and retain heat and moisture, and thereby relieve internal congestion and pain; they are sometimes employed to hasten the removal of a slough or dead tissue, and to clean the surface of foul ulcers.

Poultices are supposed to favor the development of bacteria when applied to a wound, and consequently are not used in this form of injury unless dead tissue is to be removed ; they should then be made antiseptic by sprinkling with a solution of corrosive sublimate or carbolic acid (see ANTISEPTICS). An antiseptic poultice can also be made by soaking absorbent cotton or gauze in a hot antiseptic solution and placed over the part with a covering of oiled silk ; this should be kept hot by repeated soakings.

Poultices are usually made of ground flaxseed (linseed), although, if this can not be procured, hominy, corn-meal, or bread (not the crust), may be substituted. A bread-poultice, however, becomes sour in a very short time.

Flaxseed-Poultice.—A receptacle containing boiling water should be placed on the fire. The flaxseed-meal should then be gradually added, and constantly stirred, until the batter is jelly-like, and about the consistence of oatmeal porridge. This should be evenly spread, with a thickness of about one half inch, on a cloth prepared for the purpose by folding in two or three layers, to within three inches of its borders ; over the surface of the flaxseed should be laid some antiseptic gauze, white tarlatan, cheese-cloth, mosquito-netting, veiling, or a fine cambric handkerchief ; while these fabrics do not interfere with the action of the poultice, they prevent the flaxseed from adhering to the skin. If one of the above materials can not be obtained, the surface of the poultice may be covered with a small amount of vaseline or sweet-oil as a substitute. After the poultice is applied to the body it should be covered, if possible, with oiled silk, which aids in retaining the heat; in this way a poultice can be kept warm and moist for two or three hours, or even longer. In lieu of the oiled silk, flannel or cotton wadding may be used.

A poultice can be made a deodorant, by adding to the boiling water into which the meal is thrown one or two teaspoonfuls of carbolic acid to each pint of water, or by adding to the *dry* meal, before it is thrown into the water, one half of its bulk of powdered charcoal. This mixture should be thoroughly rubbed together, and then added to the water in the manner already described. The addition of the carbolic acid makes the poultice more or less antiseptic.

Poultices made of bread, corn-meal, or hominy have nothing in their favor when compared with flaxseed, save that they may be the only available substances when needed. Poultices of these materials are made in the same manner as when flaxseed is used. A teaspoonful of salt should be added to each application, particularly when corn-meal or hominy is used, otherwise considerable irritation of the skin (covered by the poultice) may ensue.

The addition of mustard to a poultice renders it more effective in diminishing or relieving deep-seated pain, as in the chest or abdomen, and is far better, for ordinary purposes, than mustard-plasters, which are too indiscriminately used. Mustard, properly applied in this manner, is seldom followed by undue irritation of the skin. The mustard should not be added to the flaxseed in the form of powder, but should be carefully mixed with a small amount of warm water, and then stirred into the flaxseed just before it is spread upon the cloth. The amount of mustard added depends upon the degree of pain, the age of the patient, etc.; the proportion ranges from a tablespoonful to one third, or even one half, of the bulk of the poultice, one or two tablespoonfuls, however, being usually sufficient. Hot water and vinegar lessen the efficacy of mustard.

Mustard-plasters are used to obtain a rapid and decided effect, as in shock, poisoning, etc. Care should be taken, however, that the mustard does not blister the skin.

As a rule, mustard-plasters should not be applied to children and old people, as they generally blister the surface, and, if the plaster is quite large, well-marked constitutional symptoms may follow, as elevation of temperature.

In making a mustard-plaster, it is preferable to mix the powder with the white of an egg, although water may be substituted; a small amount of flour may be added, and the paste should then be thinly spread over a piece of muslin or brown paper, and covered with some thin material.

MOIST AND DRY HEAT.

MOIST HEAT.—A hot fomentation consists in the application of moist heat, and is used to relieve pain, produce relaxation, etc., and in a manner acts as a substitute for a poultice. A hot fomentation is made by wringing out of hot water a piece of flannel, old blanket, etc., which is applied immediately to the affected part and properly protected, to prevent evaporation and the consequent cooling of the dressing. For the latter reason a hot fomentation is not used when heat alone is desired. By stirring in the hot water from one to three tablespoonfuls of spirits of turpentine, the stimulant and anodyne or quieting effects of the fomentation are increased. A hot fomentation applied to the forehead and temple, when headache exists, is often followed by immediate relief.

DRY HEAT.—Dry heat is employed when the application and retention of heat are the essential elements, as in shock, syncope, drowning, etc. Dry heat is used in the form of bottles or India-rubber bags filled with hot water, or bran, salt, or sand, which has been heated in the oven and put in bags. Bran is to be preferred, as it retains heat longer, and is very light. Hot bricks, stones, flat-irons, stove-lids, plates, etc., may also be used; the latter agents should be covered with a towel or some similar substance before being applied.

In shock and some other conditions the sensation is so blunted that the skin may be burned by one of the appliances just enumerated (except the bran, salt, and sand) without being appreciated by the patient.

CHAPTER IX.

ANTISEPTICS—DISINFECTANTS.

AN antiseptic is an agent which arrests the growth of (but does not necessarily destroy) the poisonous germs (micro-organisms, bacteria) existing in the air or elsewhere, and which may infect a wound and interfere with its proper healing by producing putrefaction in the injured tissue. Blood-poisoning (pyæmia, septicæmia) may follow the absorption of these micro-organisms through an open surface. There is no opening too small to admit them.

Antiseptics are far more effective if they are applied to the wound immediately after an injury has been received. Through the recommendation of Prof. Esmarch, the German soldiers are each supplied with a small and compact package containing an antiseptic compress and bandage, and in addition a triangular bandage, upon which are illustrations describing the manner of its application on different parts of the body. The soldiers are taught that when a comrade has been wounded, the seat of injury should be immediately covered with the compress and bandages, unless the character of the wound or the hæmorrhage makes it impossible to do so. It has been undoubtedly proven that this precaution has saved the lives of many who have been wounded in battle.

It is not only important that a wound should be cleaned with an antiseptic solution, but also that it should be covered with a dressing which protects it from the air and the subsequent entrance of micro-organisms. An unclean compress introduces this form of poison, and should consequently always be avoided. Antiseptic gauze, if possible, should be selected as the proper covering of the wound. (See COMPRESSES.) This material, however, is not always available; consequently it will often be necessary to use other material, which, under the circumstances, will answer the purpose very well, provided it is *clean*, and, if possible, saturated in an antiseptic solution.

It is safer, where the facilities for cleaning the wound are inferior and the hands of the attendant are not clean, to apply the compress (whether aseptic or not) at once, leaving the wound to be made antiseptic by the surgeon.

It must not be inferred from the foregoing that a wound which is not treated by the antiseptic method is necessarily followed by serious results, for if it is properly cleaned and protected, especially if the wound be superficial, it will usually do well, cleanliness being the essential element. The antiseptic treatment, however, undoubtedly offers greater protection.

Some of the most valuable antiseptics, such as corrosive sublimate, are also disinfectants, and are among the deadliest poisons, even in very small quantities; they should therefore be used with the greatest care, and the bottle or box in which they are contained should always be appropriately and conspicuously labeled.

The number of grains or teaspoonfuls indicated in parentheses, of the following antiseptics, represent the quantity to be added to each pint of water in order to make an efficient antiseptic solution. It will be found that carbolic acid, bichloride of mercury, and some other disinfectants are not thoroughly dissolved in cold water, and produce an imperfect mixture, consequently two or three tablespoonfuls of hot water or a smaller quantity of glycerin should first be added in order to thoroughly dissolve these antiseptics, sufficient water to make the pint being added afterward.

CORROSIVE SUBLIMATE or BICHLORIDE OF MERCURY

(three grains) is the most valuable antiseptic known. It occurs in the form of small white granules. It is also very poisonous.

CARBOLIC ACID or PHENOL (two teaspoonfuls) occurs in the form of white, flaky crystals at an ordinary temperature, having the characteristic odor. Druggists usually add a small amount of water or glycerin to keep it in a liquid state; it then has the appearance of an almost colorless (sometimes reddish) oily fluid, very poisonous and corrosive.

THYMOL (ten grains).—Large white crystals, having an aromatic and pleasant odor. Not very poisonous.

OIL OF EUCALYPTUS (teaspoonful).—Colorless fluid, with an aromatic odor. Not very poisonous.

IODOFORM is a fair antiseptic, and a valuable agent for dressing indolent or unhealthy wounds. It is generally objected to on account of its unpleasant odor.

BRANDY, WHISKY, etc., contain alcohol, and when diluted with two or three parts of water may be used as an antiseptic, although not particularly effective. They are more serviceable in arresting capillary and venous hæmorrhage in a wound.

The different forms of antiseptic materials (gauze, etc.) are made by treating tarlatan, absorbent cotton, etc., with one of the above agents. It is found in the shops in compact form, and should be hermetically sealed, as in closed tin boxes, otherwise its antiseptic properties would soon become impaired.

DISINFECTANTS.

A disinfectant is an agent which has the power of destroying the infectious matter emanating from one suffering with an infectious disease, as small-pox, or from a variety of other sources.

Disinfectants are used to destroy the infectious matter contained in the apartment of the sick, and in clothing, privies, cess-pools, etc. Fumigation is the use of a disinfectant in the form of a gas, as the sulphurous acid generated by burning sulphur, or formaldehyde gas produced by the oxidation of methyl or wood alcohol.

Before describing the manner in which disinfection should be performed, it will be proper to indicate the necessary measures to be taken and precautions to be observed in a case of infectious disease. At the outset the patient should, if possible, be removed to a room at the top or a remote portion of the house in which he is dwelling.

Good ventilation, sunlight, and cleanliness are most important requisites, and are also valuable disinfectants. The room should previously be divested of everything not essential to the need of the patient, such as the carpet, drapery, unnecessary furniture, etc. This diminishes the subsequent danger, and simplifies the process of disinfection at a later period.

Communication between the patient and other members of the family (save the attendant) should be prohibited. The clothing removed from the patient, and the linen from the bed, also the dishes, etc., should be properly treated (see SPECIAL DISINFECTION). If recovery follows, the patient should be bathed daily with a warm antiseptic solution (corrosive sublimate, two grains, or carbolic acid, two teaspoonfuls, to each pint of water) for three or four days before leaving the apartment.

A dead body should be wrapped in a sheet saturated in a solution of corrosive sublimate, sixteen grains, or carbolic acid, six teaspoonfuls, to each pint of water.

The rules suggested by the Health Department of the city of New York for disinfection and fumigation are so practical and comprehensive that they are embodied in this article, and are as follow :

DISINFECTION AND DISINFECTANTS.

Sunlight, pure air, and cleanliness are always very important agents in maintaining health and in protecting the body against many forms of illness. When, however, it becomes necessary to guard against such special dangers as accumulated filth or contagious diseases, disinfection is essential. In order that disinfection shall afford complete protection, it must be thorough, and perfect cleanliness is better, even in the presence of contagious disease, than poor disinfection.

All forms of fermentation, decomposition, and putrefaction, as well as the infectious and contagious diseases, are caused by minute living germs. The object of disinfection is to kill these germs. Decomposition and putrefaction should at all times be prevented by the immediate destruction, or removal from the neighborhood of the dwelling, of all useless putrescible substances. In order that as few articles as possible shall be exposed to infection by the germs causing the contagious diseases, it is important that all articles not necessary for immediate use in the care of the sick person, especially upholstered furniture, carpets, and curtains, should be removed from the sick-room at the very beginning of the illness.

AGENTS FOR CLEANSING AND DISINFECTION.

Too much emphasis can not be placed upon the importance of sunlight, fresh air, and cleanliness, both as regards the person and the dwelling, in preserving health and protecting the body from all kinds of disease. Sunlight and fresh air should be freely admitted through open windows, and personal cleanliness should be attained by frequently washing the hands and body.

Cleanliness in dwellings, and in all places where men go, may under ordinary circumstances be well maintained by the use of the two following solutions:

1. *Soap-suds Solution.*—For simple cleansing, or for cleansing after the methods of disinfection by chemicals described below, one ounce of common soda should be added to twelve quarts of hot soap (soft soap) and water.

2. *Strong Soda Solution.*—This, which is a stronger and more effective cleansing solution, is made by dissolv-

ing one half pound of common soda in three gallons of hot water. The solution thus obtained should be applied by scrubbing with a hard brush.

When it becomes necessary to arrest putrefaction or to prevent the spread of contagious diseases by killing the living germs which cause them, more powerful agents must be employed than those required for simple cleanliness, and these are called disinfectants. The following are some of the most reliable disinfectants:

3. *Heat.*—Complete destruction by fire is the best method of disposing of infected articles of small value, but continued high temperatures not as great as that of fire will destroy all forms of life; thus, boiling or steaming in closed vessels for one half hour will destroy all disease germs.*

4. *Carbolic-acid Solution.*—Dissolve six ounces of carbolic acid in one gallon of hot water. This makes approximately a five-per-cent solution of carbolic acid, which, for many purposes, may be diluted with an equal quantity of water. The commercial colored impure carbolic acid should not be used to make this solution. Great care must be taken that the pure acid does not come in contact with the skin.

5. *Bichloride Solution* (bichloride of mercury or corrosive sublimate).—Dissolve sixty grains of pulverized corrosive sublimate and two tablespoonfuls of common salt in one gallon of hot water. This solution must be kept in glass, earthen, or wooden vessels (not in metal vessels).

The carbolic and bichloride solutions are very poisonous when taken by the mouth, but are harmless when used externally.

6. *Milk of Lime.*—This mixture is made by adding one quart of dry freshly slaked lime to four or five quarts of water. (Lime is slaked by pouring a small quantity of water on a lump of quick-lime. The lime becomes hot, crumbles, and as the slaking is completed a white powder

---

* For further information regarding this agent and also formaldehyde and sulphurous-acid gases, see page 103.

results. The powder is used to make milk of lime.) Air-slaked lime has no value as a disinfectant.

7. *Dry Chloride of Lime.*—This must be fresh and kept in closed vessels or packages. It should have the strong pungent odor of chlorine.

The proprietary disinfectants, which are so often widely advertised, and whose composition is kept secret, are relatively expensive and often unreliable and inefficient. It is important to remember that substances which destroy or disguise bad odors are not necessarily disinfectants.

NOTE.—The cost of the carbolic solution is much greater than that of the other solutions, but generally is to be much preferred. When the cost is an important element, the bichloride solution may be substituted for all purposes for which the carbolic solution is recommended, except for the disinfection of discharges, eating utensils and articles made of metal, and of clothing, bedding, etc., which is very much soiled. Its poisonous character, except for external use, must be kept constantly in mind.

METHODS OF DISINFECTION IN INFECTIOUS AND CONTAGIOUS DISEASES.

The diseases to be guarded against by disinfection are scarlet fever, measles, diphtheria, tuberculosis (consumption), small-pox, typhoid and typhus fever, yellow fever, and cholera.

1. *Hands and Person.*—Dilute the carbolic solution with an equal amount of water, or use the bichloride solution without dilution. Hands soiled in caring for persons suffering from contagious diseases, or soiled portions of the patient's body, should be immediately and thoroughly washed with one of these solutions, and then washed with soap and water. The nails should always be kept perfectly clean. Before eating, the hands should be first washed in one of the above solutions, and then thoroughly scrubbed with soap and water by means of a brush.

2. *Soiled clothing, towels, napkins, bedding, etc.*, should be immediately immersed in the carbolic solution, in the

sick-room, and soaked for twelve hours. They should then be wrung out and boiled in the soap-suds solution for one hour. Articles such as beds, woolen clothing, etc., which can not be washed, should be referred to the Health Department for disinfection or destruction.

3. *Food and Drink.*—Food thoroughly cooked and drinks that have been boiled are free from disease germs. Food and drinks, after cooking or boiling, if not immediately used, should be placed when cool in clean dishes or vessels and covered. In presence of an epidemic of cholera or typhoid fever, milk, and water used for drinking, cooking, washing dishes, etc., should be boiled before using ; and when cholera is prevalent all persons should avoid eating uncooked fruit, fresh vegetables, and ice.

4. *Discharges of all kinds, from the mouth, nose, and bowels* of patients suffering from contagious diseases, should be received into glass or earthen vessels containing the carbolic solution or milk of lime, or they should be removed on pieces of cloth, which are immediately immersed in one of these solutions. Special care should be observed to disinfect at once the vomited matter and the intestinal discharges from cholera patients, as these alone contain the dangerous germs. In typhoid fever the intestinal discharges, and in diphtheria, measles, and scarlet fever the discharges from the throat and nose, all carry infection, and should be treated in the same manner. The volume of the solution used to disinfect discharges should be at least twice as great as that of the discharge. After standing for an hour or more the disinfecting solution with the discharges may be thrown into the water-closet. Cloths, towels, napkins, bedding, or clothing soiled by the discharges must be at once placed in the carbolic solution, and the hands of the attendants disinfected, as described above. In convalescence from measles and scarlet fever the scales from the skin (peeling) are also carriers of infection. To prevent the dissemination of disease by means of these scales, the skin should be carefully washed daily

in warm soap and water. After use, the soap-suds should be thrown into the water-closet and the vessel rinsed out with a carbolic solution.

5. *The Sputum from Consumptive Patients.*—The importance of the proper disinfection of the sputum (expectoration) from consumptive patients is little understood. Consumption is a contagious disease, and is always the result of transmission from the sick to the healthy or from animals to man. The sputum contains the germs which cause the disease, and in a large proportion of cases is the source of infection. After being discharged, unless properly disposed of, it may become dry and pulverized and float in the air as dust. This dust contains the germs, and is the common cause of the disease through inhalation. In all cases, therefore, the sputum should be disinfected when discharged. It should be received into covered cups containing the carbolic or milk-of-lime solution. Handkerchiefs soiled by it should be soaked in the carbolic solution and then boiled. Dust from the walls, moldings, pictures, etc., in rooms that have been occupied by consumptive patients, contains the germs, and will produce tuberculosis in animals when used for their inoculation. Therefore rooms should be thoroughly disinfected before they are again occupied. If the sputum of all consumptive patients were destroyed at once when discharged, a large proportion of the cases of the disease would be prevented.

6. *Closets, Kitchen and Hallway Sinks, etc.*—Each time the closet is used for infected discharges, one pint of the carbolic solution should be poured into the pan (after it is emptied) and allowed to remain there. All discharges should be disinfected before being thrown into the closet. Sinks should be flushed at least once daily.

7. *Dishes, knives, forks, spoons, etc.*, used by a patient should be kept for his exclusive use, and not removed from the room. They should be washed first in the carbolic solution, then in boiling hot soap-suds, and finally

rinsed in hot water. These washing fluids should afterward be thrown into the water-closet. The remains of the patient's meals may be burned or thrown into a vessel containing the carbolic solution or milk of lime and allowed to stand for one hour before being thrown away.

8. *Rooms and their Contents.*—Rooms which have been occupied by persons suffering from contagious disease should not be again occupied until they have been thoroughly disinfected by the Health Department. For this purpose either careful fumigation with sulphur will be employed, or this combined with the following procedure: carpets, curtains, and upholstered furniture which have been soiled by discharges, or which have been exposed to infection in the room during the illness, will be removed for disinfection by steam. Woodwork, floors, and plain furniture will be thoroughly washed with the soap-suds and bichloride solutions.

9. *Rags, cloths, and articles of small value,* which have been soiled by discharges or infected in other ways, should be burned.

10. *In case of death,* the body should be completely wrapped in several thicknesses of cloth wrung out of the carbolic or bichloride solution and placed in a hermetically sealed coffin.

It is important to remember that *an abundance of fresh air, sunlight, and absolute cleanliness* not only helps protect the attendants from infection, but also aids in the recovery of the sick.

METHODS OF CLEANLINESS AND DISINFECTION TO PREVENT THE OCCURRENCE OF ILLNESS.

1. *Water-closet bowls and all receptacles for human excrement* should be kept perfectly clean by frequent flushing with a large quantity of water, and as often as necessary disinfected with the carbolic or bichloride solutions. The woodwork around and beneath them should be frequently scrubbed with the hot soap-suds solution.

2. *Sinks and the woodwork around and the floor beneath them* should be frequently and thoroughly scrubbed with the hot soap-suds solution.

3. *School sinks* should be thoroughly flushed with a large quantity of water at least twice daily, and should be carefully cleaned twice a week or oftener by scrubbing. Several quarts of the carbolic solution should be frequently thrown in the sink after it has been flushed.

4. *Cess-pools and Privy-vaults.*—An abundance of milk of lime or chloride of lime should be thrown into these daily, and their contents should be frequently removed.

5. *Cellars and rooms in cellars* are to be frequently whitewashed, and, if necessary, the floors sprinkled with dry chloride of lime. *Areas and paved yards* should be cleaned, scrubbed, and, if necessary, washed with the bichloride solution. *Street gutters and drains* should be cleaned, and when necessary sprinkled with chloride of lime or washed with milk of lime.

6. *Air-shafts* should be first cleaned thoroughly, and then whitewashed. To prevent tenants throwing garbage down air-shafts, it is advisable to put wire netting outside of windows opening on shafts. Concrete or asphalt bottoms of shafts should be cleaned and washed with the bichloride solution, or sprinkled with chloride of lime.

7. *Hydrant sinks, garbage receptacles, and garbage and oyster-shell shutes and receptacles* should be cleaned daily, and sprinkled with dry chloride of lime.

8. *Refrigerators and the surfaces around and beneath them, dumb-waiters, etc.*, may be cleaned by scrubbing them with the hot soap-suds solution.

9. *Traps.*—All traps should be flushed daily with an abundance of water. If at any time they become foul, they may be cleaned by pouring considerable quantities of the hot strong soda solution into them, followed by the carbolic solution.

10. *Urinals and the floors around and underneath*

them should be cleaned twice daily with the hot soap-suds solution, and, in addition to this, if offensive, they may be disinfected with the carbolic solution.

11. *Stable Floors and Manure-vaults.*—Stable floors should be kept clean, and occasionally washed with the hot soap-suds, or the hot strong soda solution. Powdered fresh chloride of lime may be used in manure-vaults.

12. *Vacant rooms* should be frequently aired.

13. *The woodwork in school-houses* should be scrubbed weekly with hot soap-suds. This refers to floors, doors, door-handles, and all woodwork touched by the scholars' hands.

14. *Spittoons in all public places* should be emptied daily and washed with the hot soap-suds solution, after which a small quantity of the carbolic solution or milk of lime should be put in the vessel to receive the expectoration.

15. *Elevated and Surface Cars, Ferry-boats, and Public Conveyances.*—The floors, door-handles, railings, and all parts touched by the hands of passengers should be washed frequently with the hot soap-suds solution. Slat-mats from cars, etc., should be cleaned by scrubbing with a stiff brush in the hot soap-suds solution.

THE STERILIZATION OF MILK FOR FEEDING INFANTS.

Sterilization is the process employed to destroy the germs contained in milk. Germs produce fermentation (souring), and render the milk unfit to be used as an article of food for infants. Milk, as it reaches the city, even if great care has been taken in its collection and shipment, contains germs, and these will produce fermentation, although the milk is kept on ice. Unclean vessels hasten this process. No matter how good milk may be in the morning, when comparatively fresh, toward evening, unless it has been partly or completely sterilized, it may be dangerous to an infant and may cause fatal illness, even though it still tastes sweet.

STEAM — FORMALDEHYDE — SULPHUROUS ACID (SULPHUR DIOXIDE).

At the present time steam is the most valuable and thorough disinfectant known for the treatment of clothing, bedding, and other textile fabrics. The uncertainty which has existed as to the temperature and exposure necessary to kill the micro-organisms which live in infected material has been practically removed by the result of recent experiments made at the laboratory connected with the New York Quarantine Station.* The experiments referred to consisted in the application of steam-heat in a double-jacketed steel chamber to packages of clothing, bedding, etc., containing linen disks infected with cultures of micro-organisms cultivated at this laboratory. As a result of this exhaustive work, which included numerous experiments with different degrees of temperature and exposures of different length, it was shown that a temperature of 230° F., with an exposure of fifteen minutes, will kill all known pathogenic organisms (germs which cause infectious disease), even in large bundles tightly packed, the great penetrating power of this agent being well marked. It should be borne in mind that steam destroys goods composed of leather, rubber, etc. It is unfortunate that the apparatus necessary for this purpose is so extensive and complicated that, as a rule, it can only be used by health departments and other branches of the public service.

FORMALDEHYDE GAS.—During the past year (1897) experiments made at the laboratory connected with the New York Quarantine Station—as well as at the laboratory of the New York City Health Department, the United States Marine-Hospital Service, and others—have shown the value of formaldehyde gas as a disinfectant. This agent is generated by the oxidation of ordinary wood or methyl alcohol. A description of this process and of the

* See the " American Journal of the Medical Sciences," August, 1897.

apparatus necessary for its use may be found in the "New York Medical Journal" of October 16, 1897. Although formaldehyde is not by any means as efficient and thor-

Apparatus for disinfecting by formaldehyde.

ough as steam, and although its power of penetration is not so great as this agent, it is, however, an effective disinfectant, and does not, as a rule, injure or discolor the finest fabrics—which is an important consideration. The many apparatus for supplying this gas, which during the past few months have been presented to the public, are reasonable in price, easily carried, and can be manipulated by almost any one. Therefore this agent can be used where steam chambers are not available and where economy is an object.

SULPHUROUS-ACID GAS (*sulphur dioxide*).—The oldest known disinfectant. Sulphur dioxide is generated by burning ordinary sulphur. Although we are not in possession of such definite facts relative to the value of sulphur dioxide as bacteriological work has given us in regard to steam and formaldehyde, it is certain that it is a germicidal agent of great worth. Unlike formaldehyde, it frequently changes the color, or bleaches fabrics which are submitted to its use. These, however, are usually of delicate shades and made of silk, etc.; clothing and bedding are not affected in this way, and therefore sulphur dioxide is valuable for general disinfection, particularly where steam is not available. The ordinary sulphur which is used for the manufacture of this gas is almost always available, and the method of generation is exceedingly simple. The sulphur is broken into small pieces and placed in a heavy tin or sheet-iron pan, which should be large in order that the sulphur shall be fully exposed to the air. The pan should be placed in the center of the apartment to be disinfected in such a manner that it will not be brought in direct contact with the floor or other inflammable substance. A live coal of fire may be placed in the mass of sulphur and ignite it; or a better method is to sprinkle or pour over the sulphur some alcohol, which is ignited by a match. This will cause the sulphur to burn more evenly. An ordinary and safe method is as follows: Take a wash-tub; fill it with water to the height of a brick, two or three of which have been laid in the tub; a large sheet-iron or tin pan is now placed on the bricks. In this is put the sulphur broken into small fragments and set on fire in the manner already described. The combustion which now takes place generates the gas. The water which is in the tub prevents fire which might take place in the spluttering of the sulphur, or if the pan were placed on the floor or table. After this process is well begun the door should be tightly closed before the sulphur is ignited. As a preliminary measure, it is necessary that all windows and other openings should be tightly closed, otherwise the disinfection is practically worthless. This also applies to disinfection with formaldehyde. It is well to understand that disinfection performed with gas should always have a long exposure. For this reason, the apartment under disinfection should be closed from eight to twelve hours at least; also that in consideration of the fact that gaseous disinfectants have but little penetration, and that articles to be disinfected should be opened and spread out and hung up, and not rolled into bundles.

CHAPTER X.

CONTUSIONS AND WOUNDS.

CONTUSIONS—BRUISES.

A CONTUSION is an injury inflicted upon a portion of the body by a blow from a blunt instrument, also from a fall, or severe pressure, and resulting in the laceration of blood-vessels (usually small) and other structures beneath the skin, the latter remaining unbroken.

The subcutaneous escape of blood is immediately followed by swelling and discoloration of the skin ; the color being at first black and blue or purplish, then green, yellow, and so on until the extravasated blood is removed by absorption, and the affected part regains its normal color and appearance—usually within two weeks. A "black eye" is a familiar example of a simple contusion.

In severe contusions, although the skin may be at first unbroken, the soft tissues are often so badly injured that death (gangrene) of the affected structures follows.

In contusions which extend deeply into the tissues, the discoloration may not appear for a number of days.

When a contusion is followed by very rapid and extensive swelling, in which pulsation can be detected, it indicates that a large artery has been divided.

TREATMENT.—Slight contusions need no special consideration. In those of a severer nature the treatment depends upon the time that has elapsed since the injury, and its gravity. If seen early, the indications for treatment are—
(1) to prevent the further escape of blood in the tissues ;
(2) to counteract the pain, shock, or inflammatory action

that may follow ; (3) to preserve the vitality of the part, which may be endangered in severe contusions; (4) to promote the absorption of the blood which has already escaped. The first indication can be met by the use of *hot* (not warm) or cold applications, the latter usually proving more effective—particularly if used in the form of ice broken into small pieces and placed in a rubber bag made for this purpose, or in a bladder or towel, and applied to the part and retained *only* until the hæmorrhage is controlled. Although ice is a valuable agent to check the extravasation of blood, it should be used with care, and not in all cases. In slight contusions it is particularly valuable and unattended with danger; in a severe form of contusion, however, where the vitality of the affected tissues is impaired, the use of ice, by still further reducing the vitality, may cause gangrene of the parts. Compresses soaked in dilute alcoholic solutions of whisky, brandy, cologne, arnica, camphor, etc., or, solutions containing acetate of lead (sugar of lead), carbolic acid, alum, vinegar, lemon-juice, or common salt, are also very efficacious. Pressure carefully employed, in addition to the applications just referred to, will render the treatment still more effective. Elevation of the affected part diminishes the tendency to further extravasation of blood. If shock accompanies the contusion, it is to be treated according to the directions given in another chapter.

The pain that is commonly present in a contusion is usually quieted by the applications used to fulfill the first indication. The tendency to subsequent inflammation is treated by judicious use of cold, rest, elevation, etc.

In severe contusions, where the vitality of the part is greatly impaired by the obstruction to the circulation due to the escape of blood into the tissues and consequent swelling, the temperature of the part is lowered, and cold should not be used. The local application of warmth is then indicated ; and should be applied in a dry form, as bottles filled with hot water; a bag filled with bran or

CONTUSIONS AND WOUNDS. 109

oatmeal which has been heated in the oven, or whatever form of dry heat can be easily and quickly obtained, will answer. The part should also be surrounded by woolen cloths or any fabric that will retain the heat, and if a limb is the part affected, it should be slightly elevated.

The local treatment for a contusion which has existed for some time, consists in stimulating the absorption of the extravasated blood, and may be accomplished by continuous mild pressure, gentle friction alone or combined with the stimulating solutions already mentioned, which are useful for this purpose.

WOUNDS.

A wound is an injury of the outer tissues in any part of the body, associated with more or less division of the skin and deeper soft structures, and produced by a mechanical agent.

Wounds are classified as follows: Incised, lacerated, punctured, gunshot, poisoned, and contused.

Incised wounds are made by sharp cutting instruments, as knives or razors. The edges of the wound, when applied to each other, fit accurately, and completely close the opening. Hæmorrhage constitutes one of the principal dangers of this form of injury.

Lacerated wounds are made by stones, clubs, or implements which are rough or blunt, and produce more or less destruction of tissue about the wound, the edges of which are ragged and torn. Considerable local inflammation and constitutional disturbance often follow lacerated wounds.

Punctured wounds are inflicted by bayonets, daggers, swords, arrows, or other weapons which are sharp and narrow-pointed. Although the openings are quite small, these wounds usually penetrate to a considerable depth, and may injure important blood-vessels and vital organs.

Gunshot wounds, in a general way, include all injuries resulting from the explosion of gunpowder, the direct cause being bullets, cannon-balls, and other missiles; also splinters of wood, pieces of stone, etc. The degree of danger resulting from gunshot wounds depends upon the hæmorrhage—particularly that occurring internally—the structure or organ involved, the amount of tissue destroyed, together with the shock, and the subsequent inflammation and suppuration, and blood-poisoning which may result.

The conical ball used at the present day is more dangerous than the round one. The latter is quite commonly deflected or turned aside from its original course by bony prominences, fasciæ, or tendons, and thus often prevents injury to internal organs, and when imbedded in the deeper structures is more apt to become encysted—that is, the ball is provided, by the tissues adjacent to it, with a covering or capsule, which prevents the missile from irritating the contiguous structures. Thus protected, the ball may remain indefinitely without causing harm.

Foreign substances, such as bits of clothing, are often carried into the body by bullets and other missiles (frequently beyond observation), and constitute an additional element of danger. The point of entrance of a ball is apt to be smaller than the exit, as the result of the diminished velocity.

Poisoned wounds are caused by the introduction into the tissue, through the skin, of some form of virus, as in a snake-bite. These wounds are usually punctured, although they may be lacerated, as in the bite of a rabid dog.

Contrary to the general belief, the bite of the venomous snakes in this country, such as the rattlesnake, moccasin, copperhead, and one or two others, is not generally fatal, and, although the virus acts with great rapidity and intensity, only about one out of every seven or eight bitten succumb to it. Snakes are most dangerous in warm weather and after fasting. A person who has been bitten by a poisonous serpent becomes faint and greatly depressed within a few minutes, the pulse feeble, and the pupils dilated ; more or less delirium occurs, and the extremities become cold and clammy. Considerable swelling and discoloration usually take place about the wound; intense pain is also present. In a certain proportion of cases death ensues within a few hours.

A contused wound is one in which the division of the soft structures is associated with more or less contusion at the site of injury.

HEALING OF WOUNDS. — Although the healing of wounds is divided by surgeons into a number of different varieties, it will be sufficient in this description to recognize but two of them—*union by first intention;* and *granulation,* or *union by second intention.*

Union by first intention usually occurs when the edges of the wound fit accurately, and are not displaced by hæmorrhage or improper dressing; when foreign bodies are removed and the wound properly cleaned, and the injured part kept quiet. It is always desirable, if possible, to obtain this result, which, besides other advantages, prevents disfiguring scars.

Wounds associated with considerable destruction of tissue, as in lacerated wounds, burns of the third degree, etc., heal by granulation. The first step in this form of repair is the removal by nature of the destroyed and useless tissue about the wound by the processes known as suppuration or "maturation," and sloughing. After the wound has been thus cleaned, granulation becomes apparent; little conical shoots about the size of a pin-head, and pinkish in color, are found filling up the cavity of the wound. These little bodies are very vascular, and when excessive in size, and too rapid in growth, they rise above the surrounding part, and are commonly known as "proud flesh." After granulation has entirely filled the wound, the upper surface becomes smooth, shining, and red, which appearance is the result of the process of cicatrization, and is known as a "scar" or cicatrix. The scar gradually becomes even whiter than the normal skin, and undergoes more or less contraction. The glands of the destroyed skin are not reproduced in the scar, consequently hair does not grow from the new formation.

The proper healing of wounds is often interfered with by severe local inflammation and constitutional disturbances. Surgeons believe such phenomena to be due to the existence of microscopical particles in the air, called " micro-organisms," or bacteria, which infect fresh wounds. The theory has developed within recent years a method known as the antiseptic treatment, which means the employment, about the wound, of agents which destroy or retard the growth of the micro-organisms. The success of this treatment depends not only on the application of antiseptics to the wound, but on rendering aseptic everything that comes in contact with it, even the hands of the attendant, the dressings, and instruments. Surgeons now use aseptic sutures (stitches) of different substances, most commonly those made of catgut, which are absorbed in the wound—their removal being unnecessary—a very important consideration. If drainage is essential, aseptic tubes are introduced into the wound to favor the free exit of discharges. The wound is then covered with aseptic or antiseptic gauze, secured by bandages, and left undisturbed until healing has taken place (about three days), provided the dressings do not become soiled or offensive. It is not expected that members of the ambulance corps, or other non-medical persons, will be able to comply with the details just given, but the principle should be remembered and followed as nearly as the circumstances will permit.

TREATMENT OF WOUNDS.—The treatment of a wound consists of the following indications: arrest of hæmorrhage; examination of the wound, and removal of all foreign matter therefrom; support and protection of the injured part; and rest.

The variety of hæmorrhage should be determined, whether arterial, venous, or capillary, and arrested in the manner described in the chapter devoted to that subject.

A wound should be examined as to its variety—whether incised, lacerated, etc.—and also as to the structures involved. A careful search should be made for foreign bodies, which, if allowed to remain, would interfere with the proper healing of the part. Pieces of clothing, splinters, etc., should be picked out with clean fingers, and the wound then cleansed with an antiseptic solution (see ANTISEPTICS). If this is not available, pure water, or, still better, water which has been boiled (thus destroying the germs it contains), may be used.

The further treatment depends upon the character of the wound. If incised, the edges should be brought closely and accurately together and retained in apposition, which may be done by the use of strips of adhesive plaster (a poor substitute, however, for the sutures introduced by the surgeon) and a compress, or the compress alone. If adhesive plaster is used, it should be cut into strips about one half an inch in width, and immersed in a *hot* antiseptic solution, if possible. They should not entirely surround a limb, as they would then interfere with the circulation. The strips should be applied with intervals between them, to allow the free exit of any matter or pus which may form at the site of the injury; one end of the strip being placed on one side of the wound, the edges of which are held closely together, the remainder of the strip is carried over the wound and fastened to the opposite side. In removing adhesive plaster, both ends of the strip should be loosened at the same time and carried from the skin toward the wound, thereby preventing the separation of its edges.

Compresses which support and protect the wound should be made of some clean, soft material, as linen, muslin, lint, flannel, absorbent cotton, oakum, or, best of all, antiseptic gauze, which can now be purchased in any drugstore. If the antiseptic gauze can not be procured, the material used for the compress should, if possible, be made antiseptic ; otherwise it should be applied dry, as simple water favors putrefaction. (See COMPRESSES.)

The compress should be held in place by a bandage, which also helps to keep the edges of the wound together and prevents hæmorrhage (see BANDAGES). The dressings should remain undisturbed until healing takes place, unless they become offensive, or constitutional symptoms occur.

Lacerated wounds, which are associated with more or less destruction and loss of tissue, heal by granulation; consequently no effort need be made to bring the edges in direct apposition, which might subsequently interfere with the proper escape of discharges; otherwise the same kind of dressing should be applied as in incised wounds. Lacerated wounds should be cleansed and fresh dressings applied whenever indicated by the presence of an offensive odor, or a soiled condition of the compress and bandages. Owing to the very vascular condition of the face and scalp, lacerated wounds of these parts, if not too severe, often heal by primary union, and may be treated as incised wounds until some evidence of suppuration or sloughing appears.

The *immediate* local treatment of a punctured wound is very simple, and consists in applying a soothing antiseptic lotion. The result of such a wound is extremely uncertain, and depends upon the implement causing the injury, and should be carefully watched for any evidence of subsequent inflammation. Punctured wounds made by arrows are generally serious, because the arrow-head usually remains in the tissues. In this case the arrow should, if possible, be withdrawn by a rotary movement, or, if near the surface, it can be pushed out through sound tissue, its exit being accelerated by making a small cut or incision over the arrow-point. If the shaft has been broken off, and the arrow-head remains in the tissues, it will be necessary for the surgeon to cut down and extract it.

In gunshot wounds the first effort should be to arrest hæmorrhage, to protect the wound from the air, and treat the accompanying shock. However, all foreign bodies which are about the surface of the wound and are not retained by blood-clots can be removed. No exploration for the bullet should be made except by a surgeon. The wound should, if possible, be dressed antiseptically and splints applied to prevent movement of a limb, and the patient at once conveyed to a place where he can secure the proper surgical treatment.

When a person has been bitten by a snake, the wound should be immediately applied to the mouth, and the poison removed by suction. If the wound be inaccessible to the patient himself, this operation should be performed by some one else. It is believed that the venom has no effect upon the mucous membrane of the mouth unless *cuts or abrasions are present*. If an extremity is bitten, in addition to suction, the part should be immediately surrounded by a tight bandage *between the wound and the heart*, and in this way prevent the absorption of the virus into the system. It should not be forgotten that the bandage must occasionally be loosened to prevent any serious interference with the general circulation of the part. Cauterization would be of no value in this variety of poisoned wound. The depression that follows the absorption of the poison demands the free use of stimulants—whisky or brandy; however, it is not necessary or proper that the patient should be made intoxicated; over-stimulation would produce the same effect as the poison, viz., depression. Ammonia is a very valuable remedy, and may be given in the form of the carbonate of ammonia, ten or fifteen grains in whisky every half-hour, or the spirits of ammonia (hartshorn) may be substituted; one half to one teaspoonful may be administered in the same manner (in whisky). If necessary, stimulants may be given by the rectum, but always in larger doses. A very valuable method of administering the different stimulants is that commonly used by physicians, viz., by the hypodermic syringe, and is often resorted to when a patient can not easily swallow, or where the prompt action of the remedy is demanded.

The treatment of a wound caused by the bite of a dog supposed to be suffering from hydrophobia is substantially the same as in snake-bites, although, if suction is impracticable, the wound may be cauterized by heating the blade of a penknife, button-hook, or piece of wire, red-hot and applying it to the wounded surface. The poison introduced does not develop rapidly, and the constitutional symptoms may not appear for weeks; therefore stimulants are only indicated when shock is present. The suspected animal should not be killed, but placed in confinement, and carefully watched in order to be certain whether or not he is suffering from hydrophobia.

The wounds caused by tarantulas, centipedes, spiders, bees, wasps, and other insects, are very rarely dangerous. The local application of dilute ammonia, or a solution of bicarbonate of soda, is regarded as the most effective remedy, and generally relieves the pain at once. Wet fresh earth, common salt, or a slice of an onion are also valuable, or some soothing application, such as solutions containing sugar of lead, laudanum, etc., that may be procured, can be substituted. Stimulants may be indicated in some cases.

Wounds of the abdominal walls are very dangerous, particularly so if the injury extends to the abdominal cavity (which is frequently the case), the external opening being often sufficiently large to allow of the escape of the bowels or intestines. If this occurs the mass should be covered with clean cloths, wrung out of hot water for protection until the arrival of the surgeon. In the mean time the exposed bowels should be protected by some warm and clean covering. A wound of the abdominal walls which does not enter the cavity, although dangerous, should be treated as an ordinary wound. · The shock which accompanies abdominal wounds must receive appropriate treatment.

Wounds of the thorax or chest are often associated with injury to the lung. Should this complication exist, it can be recognized by pain and irritation, coughing, difficult breathing, hæmoptysis (spitting of blood), and the appearance of blood and mucus, and sometimes air at the external opening.

When this condition is present, the external wound should be closed and a compress and bandage firmly applied, and the patient placed in a recumbent position and treated in the manner already described (see HÆMOPTYSIS). Should great distress follow, the dressing must be removed, and the patient turned on the side corresponding to the wound, thus favoring the escape of accumulated blood in the chest, which probably caused the oppression.

Rest is essential to the proper healing of wounds, and should be insisted upon, particularly if the injury is of a serious nature. The limbs may be kept quiet by the use of splints or slings.

Gangrene, or *mortification*, is a condition representing destruction or death of the soft tissues, and is analogous to *necrosis* in bone. It may affect a small part, as a crushed toe, or involve a whole extremity. The latter often happens when the main artery of the arm or leg is torn across as a result of an injury, or when this vessel is subjected to prolonged pressure; this sometimes occurs when a bandage has been too tightly applied.

Among the most prominent signs of gangrene are persistent loss of heat and sensation, pallor of the part, which later becomes dusky and mottled, purplish, and at the end almost black. Decomposition or putrefaction and the formation of offensive gases follow. The presence of the gases gives a crackling sensation to the touch.

The condition just described is known as *moist gangrene*, and is the form usually met with. This term is used to distinguish it from a less common variety, known as *dry gangrene*, or *mummification*. This occurs in very old people, where the arteries are more or less diseased and the circulation is weak. It usually affects the toes, and develops very slowly. As the name implies, the affected tissue has the appearance of being dried and shriveled.

A bedsore is a localized form of gangrene, appearing about the buttock, along the spine, shoulders, and elbows, caused by long-continued pressure on these prominences in invalids who are confined to the bed, and who are often in an emaciated, enfeebled, or paralyzed condition.

In order to prevent the formation of bedsores, strict cleanliness should be observed: the bed must be soft, and smoothly made; pressure on the exposed parts must be avoided. An air-cushion beneath the sheets is a valuable means of preventing pressure, or, if possible, the patient should at certain intervals change his position in the bed, or, best of all, be placed on a water-bed. Feather beds should always be avoided for the sick. The skin over the exposed part should be hardened by bathing the exposed part twice daily with cologne water, alcohol, etc.

Treatment.—Gangrene represents dead tissue, and must be removed as early as possible.

Gangrenous extremities are amputated by the surgeon. In these cases a fatal result often follows, owing to the absorption of septic or poisonous matter into the system.

Small gangrenous spots, such as follow the lesser injuries, and also bedsores, are usually left to Nature to remove the dead tissue. This is done by a process known as "sloughing." The treatment during this period should be directed toward assisting Nature, by the use of warm applications such as carbolized flaxseed poultices, which facilitate the removal of the "slough," and by cleaning out the discharge with warm antiseptic solutions. The latter agents also diminish the offensive odor which is always present. When the slough or gangrenous matter has been cleared away the granulation and healing should be stimulated by the application of carbolic or oxide-of-zinc ointment, balsam of Peru, iodoform, etc., and the appropriate dressing applied.

Dry or senile gangrene is a slow process, and the treatment should be left with the surgeon, if possible; if not, the following simple measures may be observed. In the early stage, when the big toe (the common seat of the affection) is first involved, the part should be enveloped in cotton batting to preserve an even temperature, and the foot should be kept quiet; later, when decomposition occurs, antiseptic solutions should be applied.

CHAPTER XI.

HÆMORRHAGE.

HÆMORRHAGE is the escape of blood as the result of an injury to a blood-vessel, and is classified as *arterial, venous,* and *capillary.*

In arterial hæmorrhage the blood is thrown from the injured vessel in jets or spurts, and has a bright-red or scarlet color.

In venous hæmorrhage the blood flows from the wound in a slow, steady stream, the color being dark red or purple.

In capillary hæmorrhage the blood oozes from the general surface of the wound and not from one point, as in arterial or venous hæmorrhage—the color being dark red.

Hæmorrhage is arrested in two ways—by *natural,* and by *artificial* means.

THE NATURAL MEANS.—The natural means of arresting *arterial* hæmorrhage is as follows : After an artery has been entirely divided, its muscular coat produces a contraction and retraction of the vessel at the seat of injury. The contraction diminishes the diameter of the artery, while the retraction draws the end of the divided vessel backward into its sheath. The blood at the mouth of the bleeding vessel forms a coagulum or "clot," which aids in preventing the further escape of blood. Exposure of the bleeding surface to the air, and in severe hæmorrhage the weakened force of the heart and circulation, which often results in syncope or fainting, greatly favor the formation of the clot, and consequently help to arrest the hæmorrhage. The clot at the mouth of the injured vessel subsequently becomes organized and firmly attached, and permanently stops the bleeding. When an artery is only partly divided, the contraction and retraction can not properly take place, and the hæmorrhage is very persistent. If an artery is severed by a dull or rough instrument, or torn across, the fibers at the end of the vessel being ragged and uneven, more effectually close the opening and assist in the formation of a clot, and may arrest hæmorrhage even in large arteries. This is illustrated in machinery accidents, where an arm has been torn from the body and followed by very little or no bleeding.

The manner in which nature arrests *venous* hæmorrhage is by the contraction and retraction of the end of the bleeding vessel, with the formation of a clot, as above detailed in arterial hæmorrhage, and also the collapse of the vein at the point of injury (see VEINS). The blood-pressure in veins being much less than that of arteries, the hæmorrhage is less vigorous and more easily controlled.

In *capillary* hæmorrhage the minute size of the vessels and the contraction that follows their division, together with the rapid formation of a clot, particularly when the bleeding surface is exposed to the air, usually check the hæmorrhage in a very short time.

In some persons a condition of the blood exists which retards or prevents the formation of a clot; for this reason the most trivial wound may be followed by a hæmorrhage which can not be arrested, and terminates fatally. These subjects are known as *bleeders*. Cases of this kind are fortunately rare.

THE ARTIFICIAL MEANS.—The artificial means of arresting hæmorrhage are as follows : Position, pressure, cold and heat, torsion, rest, styptics, and ligation.

Treatment of Arterial Hæmorrhage. Position.—Elevation of the injured part aids in controlling arterial hæmorrhage only where very small vessels are involved, but is of very little value where large ones are implicated, owing to the strong pressure of blood in the arteries.

Pressure.—This method of arresting hæmorrhage is, next to ligation, the most important means that can be employed for this purpose. It may be applied directly to the bleeding surface, or along the course of the artery. In the latter instance the pressure should be always made *between the wound and the heart*. In the first variety of pressure the finger (digital pressure), or a pledget or tampon (see TAMPONS), is pressed into the wound, and constitutes a valuable means of checking the flow of blood, particularly from small arteries. Pressure with the finger can only be of avail for a short time, unless the one making the pressure can be relieved at brief intervals. For this reason, a tampon—if properly applied—is more effective than digital pressure, provided the surgeon is not expected for some time, or in case it is necessary to move the patient.

In tamponing, the wound should be thoroughly filled from the *bottom*. A tampon improperly applied is worse than useless. The material used for this purpose must be absolutely clean, and if possible made antiseptic (see ANTISEPTICS). Should the surroundings permit, choice should be made of one of the following : small strips of antiseptic gauze, linen, or muslin, or absorbent cotton. After the above has been complied with, the tampon should be held firmly in place by a bandage (see BANDAGES), but should not be allowed to remain longer than twelve or fifteen hours without the advice of a surgeon. If the hæmorrhage recurs at the expiration of this time, it may be necessary to clean the wound and apply a fresh tampon, otherwise it should receive the attention of an ordinary wound.

The second variety of pressure (along the course of an artery) is used particularly in hæmorrhage from large arteries, and is most serviceable when the vessel can be pressed against a bone. Either the finger (digital pressure), or a pad or compress (see COMPRESSES), may be used for this purpose.

If the surgeon is expected to arrive in a short time, digital pressure is the simplest, quickest, and best means that can be employed. However, if considerable time must elapse before the patient can receive the proper surgical attention, or if he is to be removed some distance, the pad should be utilized.

In digital pressure, the thumb should preferably be used.

The function of a pad applied over the course of a bleeding artery is similar to that of a tourniquet. A tourniquet is an instrument used by surgeons to arrest arterial hæmorrhage by compression.

An excellent tourniquet can be extemporized by folding a large handkerchief in the form of a cravat, placing between the folds a smooth stone, piece of wood, cork, potato, etc., or a good-sized knot may be tied in the handkerchief; the latter, however, is inferior to the stone, etc. The handkerchief is then bound *loosely* around the limb and tied, the portion acting as the pad being placed over the artery, *between the wound and the heart*, and held securely in this position, while a bayonet, sword, cane, umbrella, stick of wood, etc., should be passed between the handkerchief and the skin at the side of the limb opposite the pad and twisted until the hæmorrhage ceases (Fig. 72). As the constriction resulting from this form of pressure interferes with the return or venous circulation, the handkerchief should be occasionally loosened if there is evidence of serious obstruction, as swelling and blueness of the part below the constriction; digital pressure should be substituted while the handkerchief is loose. A handkerchief is given as an illustration of what may be at once secured, although parts of clothing, neck-ties, suspenders or rubber tubing, rope, etc., are equally valuable, provided they fulfill the requirements. Hæmorrhage may be checked without the pad by simply constricting the part with one of the agents just enumerated. This, however, is less effective.

Prof. Esmarch has recently devised a suspender made of elastic webbing, and so constructed that, after detaching it from the trousers, the apparatus (straps, etc.), by which it is attached, can be easily and quickly removed, leaving a long elastic ribbon which is to be used in constricting a bleeding limb (Fig. 65).

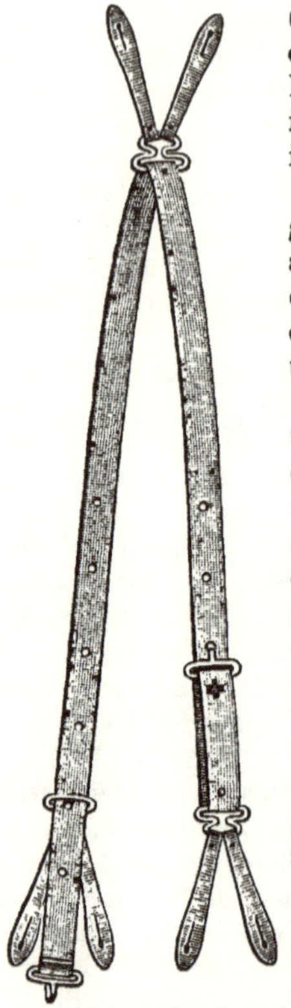

Torsion is a method used by surgeons to check bleeding from small arteries. It consists in catching the end of the injured vessel with forceps or clamp and twisting it until the hæmorrhage ceases.

Ligation or ligaturing (tying) is the procedure by which the end of the bleeding vessel is tightly constricted and securely closed by an agent called a ligature, which consists of a thread of silk, catgut, etc. A ligature is commonly employed to arrest hæmorrhage (particularly in arteries), and is the most valuable method of accomplishing this result. The use of this means, however, should remain within the province of the surgeon.

Treatment of Venous Hæmorrhage. Venous hæmorrhage is easily controlled if the following rules are observed:

Fig. 65.—Suspender devised by Prof. Esmarch.

1. Remove everything *between the wound and the heart* that retards the flow of blood, as garters, tight clothing, etc.
2. Elevate the injured part.
3. Apply a good firm compress directly to the wound.

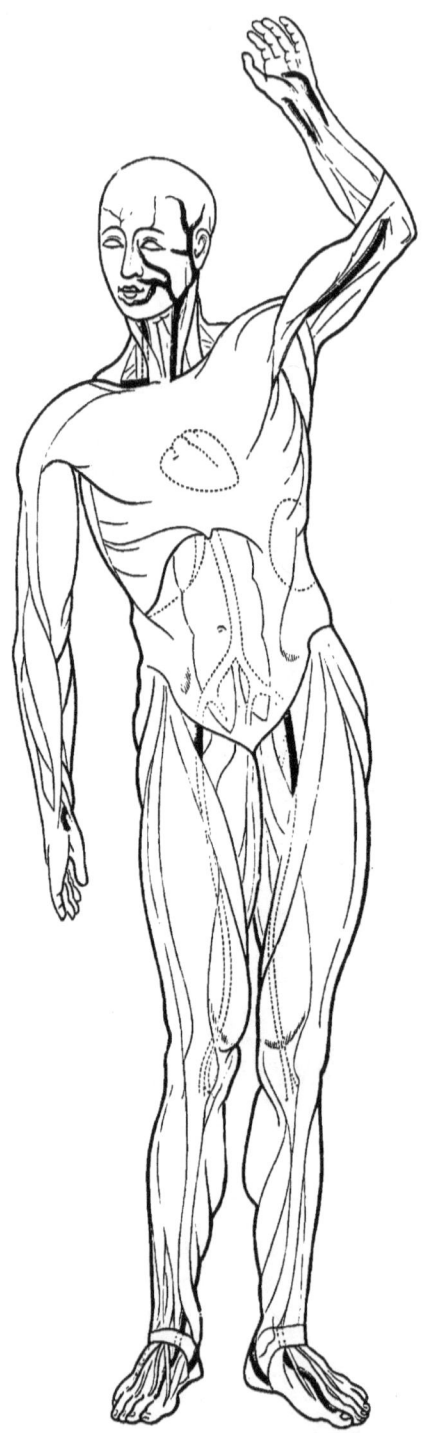

Fig. 66.—Diagram showing the position of the important arteries.

Constriction of the limb *beyond or below* the seat of hæmorrhage is valuable, but inferior to the method just described.

The importance of the first and second rules will be apparent in rupture of varicose veins in the leg, which often bleed freely; the valves of these disabled veins being rendered useless, the blood escapes from both ends of the divided vessel. In rupture of varicose veins of the leg a bandage should be applied over the compress, beginning at the toes and extending upward to a short distance above the seat of hæmorrhage. In ordinary venous hæmorrhage the free return of blood to the right side of the heart, aided by elevation of the limb, relieves the blood-pressure in the veins in the immediate vicinity of the wound, and consequently the hæmorrhage is not so profuse.

Treatment of Capillary Hæmorrhage. Capillary hæmorrhage is usually harmless, except in case of "bleeders," and generally ceases when the bleeding surface is exposed to the air. If such means are not successful, hot or cold applications or a compress should be employed.

Cold is a very valuable means of controlling venous and capillary hæmorrhage, and *aids* in arresting the bleeding from an artery. It may be applied in the form of cold air (exposure of the bleeding surface), cold water, ice, and snow.

Hot applications (temperature 120° to 125° Fahr.) are superior to cold. A piece of flannel wrung out in water as hot as can be borne by the skin, and applied directly to the bleeding surface, is followed by a diminution or a cessation of hæmorrhage. Both hot and cold applications contract the bleeding vessels and hasten the formation of the clot.

The application of alcoholic solutions, or the spirits of turpentine, to the wound by a cloth saturated with either, although very irritating, are sometimes used to check hæmorrhage.

Styptics, or astringents, although powerful agents for arresting hæmorrhage, are used by surgeons with great reluctance and only in special cases, or where other means are not available or less effective. Styptics are objectionable, for the reason that the stronger ones, as solution of the subsulphate of iron (Monsel's solution), nitrate of silver, and some others, injure the wounded surface, and may be followed by sloughing. The weaker styptics or astringents, such as tannin, gallic acid, matico, cobwebs, and alum, prevent healing by first intention. Alum is the least objectionable of the latter group. The application of styptics to mucous membranes (mouth, nose, etc.) is followed by more favorable results than when applied to a raw, wounded surface.

Rest is extremely important in all varieties of hæmorrhage, as it favors the formation and retention of the clot.

Hæmorrhage occurring in the scalp may be easily arrested by the use of a compress and bandage (see BANDAGES), which press the bleeding vessels against the underlying skull.

Hæmorrhage from the mouth may usually be stopped by the use of ice and astringent (alum, tannin) or alcoholic (brandy, whisky, etc.) solutions; if not sufficient, a tampon should be firmly held against the bleeding point. In severe cases the common carotid artery (see description of this vessel), on the side corresponding to the injury, may be compressed, although this should only be used as a last resort.

The lips are supplied by arteries which divide at the angles of the mouth, and entirely surround this opening. When the lips have been injured, the hæmorrhage may be checked by pressing the sides of the wound between the thumbs and fingers.

Severe hæmorrhage following the extraction of a tooth can be controlled by replacing the tooth, or by the application of a tampon saturated with a strong solution of alum, or other astringent, to the cavity.

Epistaxis, or "nose-bleed," is the most frequent form of internal hæmorrhage, and may be controlled in the following manner: Elevation of the head and arms, removal of all constriction about the neck, cold applications to the back of the neck, forehead, or bridge of the nose. Passing two fingers beneath the upper lip and directing pressure against the base of the nose or nostrils is very useful; snuffing up some finely powdered tannin or cold, salt, or alum water, or, even better, syringing the nasal cavity with some of these remedies. If the mouth is kept open during this operation, the fluid will escape through the opposite nostril. A very valuable method of checking epistaxis is by the use of a thin rubber finger-cot or protector, which should be greased and carefully passed into the nasal cavity of the affected side; after the cot is in position it should be filled or partly filled with very small pieces of ice. In severe cases of epistaxis which resist the remedies already enumerated, the nasal cavity on the side corresponding to the hæmorrhage may be tamponed by carrying into the nasal cavity of the affected side a piece of gauze (antiseptic, if possible) over a pencil or similar agent. When the pencil is withdrawn the cavity left in the gauze is packed with small pieces of cloth or cotton. The tampon should be left a number of hours, and never forcibly removed, but should be loosened by injection of water or oil.

In order to be able to properly control hæmorrhage by pressure, it will be necessary to know the position of certain arteries and their relation to contiguous structures, so that they may be readily found and compressed. For this purpose the following diagrams have been introduced, which will afford a guide in locating these blood-vessels.

Common Carotid Artery.—This vessel supplies the head with arterial blood. Its course from below upward corresponds to a line drawn from the junction of the collar and breast bones (clavicle and sternum) upward to a point just behind the angle of the lower jaw, or between it and the mastoid process of the skull—a bony prominence just back of the ear; this line also indicates very closely the anterior or front border of the sterno-mastoid muscle, which extends from the mastoid process to the sternal end of the clavicle already mentioned.

The application of digital pressure to the common carotid artery is indicated in severe hæmorrhage about the

FIG. 67.—Digital compression of common carotid artery.

head and upper part of the neck (as in "cut throat"), which can not be controlled by other means. The pressure should be applied about midway in the neck at the anterior border of the sterno-mastoid muscle, and directed against the anterior portion of the spinal column in the neck (Fig. 67). The common carotid artery is accompanied by a large vein (internal jugular) and a very important nerve (pneumogastric) which may be injured if the part be roughly manipulated.

In wounds of the hand, forearm, and arm, associated with severe arterial hæmorrhage, pressure may be made upon the subclavian, axillary, brachial, or radial and ulnar arteries, according to the situation of the wound.

Subclavian Artery.—Pressure should be applied to

this vessel in hæmorrhage occurring at the upper part of the arm or in the axillary space ("arm-pit"). The outer

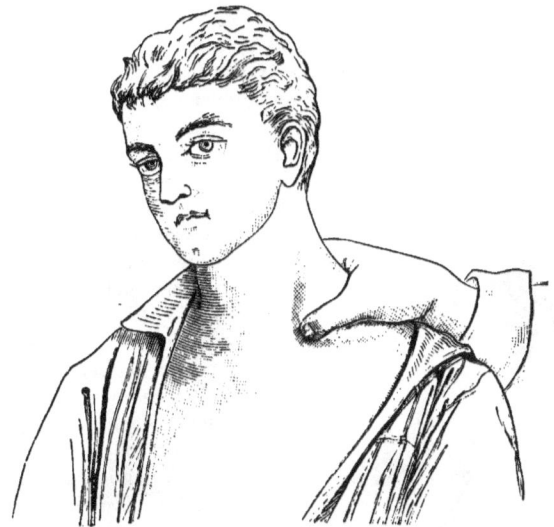

FIG. 68.—Digital compression of subclavian artery.

portion of the subclavian artery passes over the upper surface of the first rib. If the thumb is directed downward beind the clavicle, about two inches from the breast-bone, the artery may be reached and compressed against the first rib (Fig. 68). Pushing the shoulder of the patient downward facilitates this procedure. In some persons digital pressure fails to arrest the hæmorrhage, as the artery can not be reached by the finger; then the handle of a door-key or some other agent suitable for the purpose may be substituted.

Axillary Artery. — This vessel is the continuation downward of the subclavian artery. It passes through the axillary space, and can not be easily compressed in this situation; consequently, the subclavian is generally selected as the artery to which pressure is to be applied in hæmorrhage occurring in the upper part of the arm. Some pressure may be made upon the axillary artery by raising the patient's arm and pressing the vessel against the upper and inner portion of the humerus. A book, or whatever may be used, should be carried well up into the axillary space, and the arm brought close to the side. It should be remembered that undue pressure at this point may injure some of the numerous nerves in this vicinity.

FIG. 69.—Line showing the course of the brachial artery.

Brachial Artery.—Owing to the numerous injuries of the upper extremity, the brachial artery requires compression oftener than any other vessel in the body. Pressure upon the brachial artery is easily and accurately made. Its course is along the inner side of the biceps muscle, which is found in the front of the arm and stands out very prominently, particularly in muscular subjects (Fig. 69). Pressure should be applied at the inner border of the biceps and directed against the humerus (Fig. 70). If the patient is to be removed, a tourniquet should be made and adjusted in the manner already described, or two round pieces of wood about the diameter of a broomstick, and properly covered to prevent injury to the skin, may be used to compress the brachial artery. One piece should be placed transversely on the inner side of the arm against the artery, while the other occupies a corresponding position on the outer side of the arm; the ends of the sticks are then tied together with sufficient force to arrest the hæmorrhage.

Hæmorrhage in the forearm and hand may be controlled by placing a small pad at the bend of the elbow, and then flexing (bending) the forearm upon the arm and keeping it in this position.

The *radial* and *ulnar* arteries are branches of the brachial, and continue down the forearm to the hand; they are superficial at the wrist, the only point at which they can be compressed with any degree of success. The *radial* artery or "pulse" may be found about three quarters of an inch from the outer border of the wrist (when the palm of the hand is turned upward), the *ulnar* being about one half inch from the inner border of the wrist. These vessels may, to a certain extent, be compressed, but in wounds of the hand associated with hæmorrhage pressure upon the brachial artery should be employed.

FIG. 70.—Digital compression of brachial artery.

A very persistent form of hæmorrhage sometimes follows a wound of the palm of the hand. The bleeding should be checked by plugging the wound and then placing in the palm of the injured hand *over* the tampon a smooth and round piece of wood, potato, apple, lemon, billiard-ball, or whatever may be closely grasped. The hand should then be bandaged in this position. The hæmorrhage may be greatly diminished or even controlled by having the patient hold tightly one of the agents just mentioned, without the addition of the tampon.

Femoral Artery.—After the external iliac artery (in the abdominal cavity) passes into the leg, it is known as the femoral artery and supplies the lower extremity. It should be compressed in severe hæmorrhage from this portion of the body. The course of this artery is from the middle of the groin downward to the inner side of the knee (Fig. 71). The artery is superficial in the upper part of the thigh, from the groin downward about six or eight inches, and it is in this situation that the pressure must be applied (Figs. 72 and 73). It is compressed most effectively where it crosses the pelvic bone at the "groin."

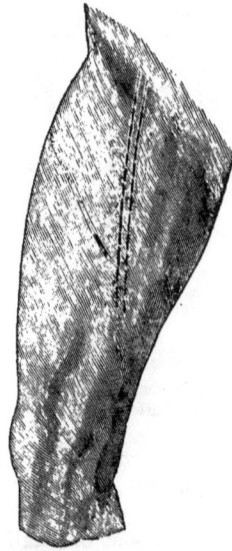

FIG. 71.—Line showing course of the femoral artery.

Popliteal Artery.—This vessel is the continuation of the femoral artery, and is found in the popliteal space (behind the bend of the knee), and is only slightly affected by digital pressure, although a pad can be placed in the popliteal space and pressed upon the artery by flexing the leg upon the thigh and securing it in this position.

Hæmorrhage from the sole of the foot is sometimes controlled by compressing the *posterior tibial*, a branch of the popliteal artery, as it winds around the ankle between the internal malleolus (ankle) or lower end of the tibia, and the heel, rather closer to the former. Plugging is also valuable for this form of hæmorrhage.

Secondary hæmorrhage means the reappearance of

bleeding in the wounded part soon after it has been once arrested. It is generally capillary, and is caused by the reaction and increased force of the heart, the relaxation of the bloodvessels at the seat of injury, or the immoderate use of stimulants, undue movement of the part, and also increased warmth to the surface that follows when the patient is placed in bed.

Elevation of the injured part and moderately increased pressure are often all that need be done, although, if the hæmorrhage continues, the dressings should be removed, the blood-clots cleaned out of the wound, and fresh dressings applied. Secondary hæmorrhage may also occur, when sloughing is present, some days after the wound is received.

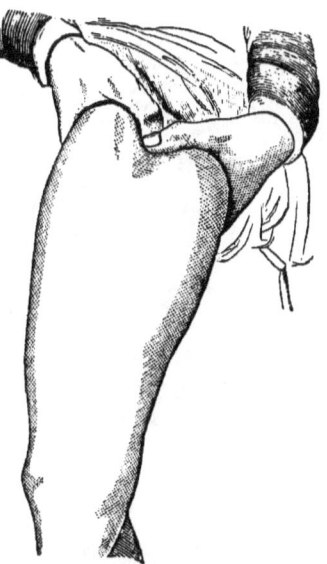

FIG. 72.—Digital compression of the femoral artery.

It should be well borne in mind that secondary hæmorrhage may be followed by collapse and death.

Clots are not to be removed from a wound unless the means of cleaning and dressing the wound are at hand; they act as a temporary compress and hæmostatic.

The *constitutional symptoms* of hæmorrhage following a great loss of blood are : Pallor of the face, lips, and surface of the body, the skin being often covered with a cold sweat : the features are pinched. Thirst, shortness of breath, restlessness and sighing, vomiting, and disturbance of the functions of the brain, are more or less prominent; dimness of vision, ringing in the ears, delirium, and unconsciousness or syncope, convulsions, and death may also occur. The severity of these symptoms depends upon the amount of hæmorrhage.

The *treatment*, besides checking the hæmorrhage, consists in the careful internal use of stimulants, application of warmth to the body, hot rectal injections, and the general treatment of shock.

Hæmoptysis, or hæmorrhage from the lungs, is generally recognized by the expectoration of bright-red and frothy blood, also coughing, and pain and rattling in the chest. The usual cause is disease of the lungs, although it may follow a wound of these organs.

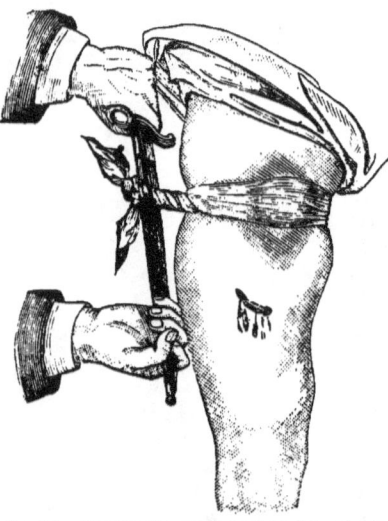

FIG. 73.—Arresting hæmorrhage from femoral artery by the use of an improvised tourniquet.

Treatment.—The fears of the patient should be calmed, and rest in the recumbent position, with the head and shoulder slightly elevated, insisted upon. The temperature of the room should be cool, and the air pure. Among the remedies commonly used are table salt, half a teaspoonful; cracked ice; ten to sixty drops of the fluid extract of ergot every hour for two or three hours, and then discontinued or taken at longer intervals; five to twenty drops of the oil of turpentine on sugar every half hour or hour until three or four doses have been taken, or half a teaspoonful of aromatic sulphuric acid (elixir of vitriol), in water occasionally. Hæmoptysis is rarely followed by a fatal termination, unless it results from a wound of the lung.

Hæmatemesis, or hæmorrhage into the stomach and vomiting of blood, is generally the result of some chronic disease of the stomach, although it may follow a blow or stab of the abdomen. The symptoms are frequently those of profuse hæmorrhage (see CONSTITUTIONAL SYMPTOMS OF HÆMORRHAGE), also a sense of fullness about the stomach, commonly followed by the vomiting of dark or black blood, which is heavy, not frothy, as in hæmoptysis. The vomited matter is mixed with food, provided the hæmorrhage occurs soon after a meal. It should be remembered that the source of hæmorrhage in hæmatemesis may be from the mouth, nose, or throat, the blood having been swallowed, consequently these parts should be examined.

The usual treatment is rest in the recumbent position; small pieces of ice should be freely swallowed, and the application of ice wrapped in a towel or in an ice-bag, or snow, cold water, etc., over the stomach; hot applications may be applied to the extremities. No attempt should be made to administer medicine or stimulants by the mouth. In severe cases of hæmoptysis and hæmatemesis the general treatment of shock is called for. Stimulants should be used with caution, as they are apt to encourage the hæmorrhage.

CHAPTER XII.

FRACTURES.

A FRACTURE is a breaking or solution of continuity in a bone.

Fractures are generally classified as simple, compound, comminuted, multiple, and complicated.

In a simple fracture the bone is broken into two fragments, but does not communicate with the outer world, the skin being uninjured.

In a compound fracture the bone is exposed to, or communicates with, the air by a wound of the soft structures.

In a comminuted fracture the bone is broken or crushed into a number of pieces at the same point, *and communicating with each other*.

In a multiple fracture the bone is broken into a number of pieces, but at different parts of the bone, and not communicating with each other.

In a complicated fracture there is, in addition to the breaking of a bone, an injury to some important adjacent structure resulting from the fracture, as blood-vessels, nerves, or joints.

A fracture is either complete or incomplete. A complete fracture is the usual variety, and involves the entire separation or loss of continuity of a bone. An incomplete fracture frequently occurs in children, owing to the elasticity of bone in early life, and has received the name of "green-stick" fracture.

The direction in which the bone is fractured is indicated by the terms transverse, oblique, and longitudinal.

An impacted fracture occurs when the broken ends of a bone are driven into each other, and remain thus fixed.

The *sign* and *symptoms* of a fracture are deformity, abnormal or increased mobility, bony crepitus, pain, loss of function, and subsequent swelling and discoloration.

The deformity is caused by the displacement of the broken ends of the bone, as the result of the violence causing the injury, also the vigorous muscular contraction at the affected part, and attempted movement on the part of the patient, which cause a shortening and change in the direction of the limb, and considerable deformity at the seat of injury. The deformity, as a rule, is not particularly apparent in impacted fractures.

Abnormal mobility is the result of the solution of continuity or break in the bone, producing a "false point of motion," which is detected by the surgeon while manipulating the part.

Crepitus, or grating, is caused by the rubbing together of the broken ends of the bone, and, when detected, is a positive sign of fracture; it is absent in impacted fracture, and also where muscular or other tissues have fallen between the ends of the broken bone; consequently, *an absence of crepitus does not mean an absence of fracture.*

Pain is caused by the contact of the fragments of bone with the adjacent structures, also the strong muscular contraction that occurs at the seat of injury. More or less heat, redness, swelling, and discoloration of the part may be present.

Loss of function is the inability on the part of the patient to make use of the affected limb.

The swelling and discoloration are due to the subcutaneous escape of blood and serum at the seat of fracture.

The repair, union, or "knitting" of the bone is begun by Nature soon after the occurrence of a fracture, and is accomplished by a substance formed at the seat of injury, known as callus, which is thrown around and between the ends of the broken bone. Although soft at first, the callus gradually hardens, and at the end of varying periods, depending on the bone injured, but not usually exceeding six weeks, the fragments are firmly united. The minute structure of callus becomes in time (about one year) similar to that of bone.

Sometimes the formation of callus is imperfect or insufficient, and the broken bone does not become united. This condition constitutes an ununited fracture.

The union of bone in a simple fracture is likened to healing by first intention in wounds, while the repair in a compound fracture resembles healing by granulation. The popular belief that the pain is increased at the seat of fracture during the "knitting" or healing process is without any foundation.

TREATMENT.—The object of a surgeon in treating a fracture is simply to assist Nature. He first carefully reduces the fracture, or "sets" the bone; that is, he endeavors, as far as possible, to bring the broken extremities in apposition or directly against each other, and by retaining them in position for a certain length of time by splints or some other form of support, the permanent union is effected by the callus, and the function of the part generally restored.

Although it is best that a fracture should be reduced and the proper dressing applied as quickly as possible after the accident, it should be remembered that the union of the fragments does not begin for some time after the injury, and that a fracture may remain several days before being reduced, and be still followed by excellent results; and also that a frequent cause of compound fracture is the outcome of unskillful manipulation. Consequently, when one not a surgeon is called upon to attend a person where a fracture is suspected, his duty consists in protecting and making immovable the injured part, and conveying the subject to a hospital, or wherever he can receive the necessary and proper treatment. However, should this be impossible, as the result of the accident having occurred where the professional services can not be procured for an indefinite period, an effort may then be made to reduce the fracture, but the manipulations are to be made only with the greatest care.

As a rule, an injured person should not be removed from the position in which he is found until an examination has been made as to the character of the injury. A violation of this rule is a frequent cause of a compound fracture.

A brief account of the manner in which the injury was received should be obtained from the patient or a bystander. An examination should then be made. If the injury is about the ankle or wrist, it can easily be exposed; however, if the affected part is nearer to the body, the clothing should be *cut* away, and not removed in the ordinary manner, which would be likely to disturb the fragments and increase the suffering. The different garments need not be indiscriminately cut, but, if possible, ripped at the seams.

If a fracture has occurred, an examination will probably show one or more of the ordinary symptoms and signs already described. If the necessary surgical attendance can be secured within a number of hours, the splints should be applied without an attempt being made to reduce the fracture ; otherwise, an effort in this direction is justifiable.

The reduction of a fracture consists in bringing the ends of the broken bone together, and is accomplished by extension and counter-extension. The term extension, when applied to the treatment of a fracture, indicates a procedure whereby the broken limb below the seat of fracture is pulled *from* the body. In counter-extension, the upper fragment, or the portion of the broken bone nearest the body, is held securely in position, or is carried in an opposite direction from the lower fragment. This manner of reducing a fracture is performed by the surgeon with the hands or an instrument devised for the purpose. The hands are generally used.

The extension and counter-extension should be made in a straight line; that is, in the long axis of the broken bone.

After the seat of injury has been examined, and the presence of a fracture ascertained or suspected, the clothing previously turned aside can now be replaced and wrapped around the part, thus affording considerable protection to it. The splint should then be adjusted, and the patient removed to a place where he can receive the proper surgical treatment.

The slings necessary to support an injured arm have already been described (see TRIANGULAR and CRAVAT BANDAGES). In addition to these, the skirt of the coat may be utilized for this purpose (Fig. 74); also other devices, such as pinning the sleeve to the waist, etc.

If the fracture is compound, no effort should be made to apply splints until the wound has been covered as quickly as possible with an antiseptic material or some form of clean dressing. When a blood-clot fills the wound, it should under no circumstances be disturbed until the patient is in the hands of the surgeon. The protection afforded by the compress or blood-clot just alluded to, prevents the entrance of poisonous germs into the system. If a point of the broken bone is pushed through the skin, it should not be interfered with, but dressed antiseptically.

FIG. 74.—Skirt of coat used as a sling.

SPLINTS.—Splints can be made of any material which is capable of rendering the part practically immovable without injuring the soft structures to which they are applied. They should be long enough to extend some distance above and below the injury, and generally including the nearest joint, and at times two or three joints, as in fracture of the thigh. Their diameter should exceed that of the limb to which they are applied, although this is not absolutely necessary, as a cane or a sword makes a very fair support. Two splints are generally used, one for the inner and one for the outer side of the limb. Splints should always be padded on the side next to the skin with some soft material, so as to prevent undue pressure and injury. After a splint has been fitted to the limb, it should be retained by the necessary bandages; the latter, however, should not surround the limb at the point of fracture, nor should they be drawn sufficiently tight to increase the suffering of the patient. This should be particularly observed in *unreduced* fractures.

Although thin wooden boards are regarded as the most desirable material for splints (at least for temporary use), being light, and easily formed to suit special cases, other substances may be employed with good results. Among those which can be secured in emergencies are shingles, laths, fence-boards, cigar-boxes, barrel-staves, bark and branches of trees, the latter being bound together with cord or green twigs. Binders' board, cut into the proper shape, makes an excellent splint. Book-covers may be used; also sole-leather, newspapers tightly and thickly folded, hay or straw bound in the same manner as the branches of trees; canes, umbrellas, broomsticks, coat- or shirt-sleeves or boot-legs stuffed with hay, grass, leaves, etc., can be utilized. A pillow, or an article of clothing properly folded, makes a very valuable temporary splint and pad combined, and is particularly useful in fracture of the leg (Fig. 79).

In military service, in addition to the above articles, guns, swords, scabbards, bayonets, leather from a saddle, etc., can be made available.

PADDING.—For padding, cotton or any soft substance,

as an article of clothing, oakum, furniture-stuffing, straw, hay, moss, grass, leaves, etc., may be used.

Bandages for the purpose of retaining the splint may be formed of handkerchiefs, neck-ties, suspenders, strips of clothing, straps, green twigs, rope, cord, wire, etc. Care should be taken that when using such substances which are small, as wire and cord, that the skin be well protected, so that it shall not be injured or cut.

Special fractures will now be considered.

FRACTURES OF THE CRANIAL BONES are usually followed by symptoms of concussion and compression of the brain. A fracture occurring at the base of the skull is usually caused by a blow about the forehead or opposite the point of fracture, or by a fall from a height, the person striking on the head or upon the feet or buttocks, and has, in addition, special symptoms which point directly to this form of injury, viz., an escape of blood from the nose and ear, and beneath the thin membranes covering the eye ; or, what is still more positive, an escape of a colorless fluid from the ear.

Treatment.—The patient should be placed on his back in a cool, dark room, and kept perfectly quiet. Cold, in some form, should be applied to the head, to prevent excessive reaction. For the same reason, the internal use of stimulants is to be avoided.

FRACTURE OF THE INFERIOR MAXILLARY BONE (lower jaw) is caused by kicks, blows, or falls. The body of the bone (the portion into which the teeth are inserted) is the usual seat of fracture, which is generally compound, having a communication with the cavity of the mouth. The deformity is shown by the irregular line of the teeth on the affected side. Crepitus, swelling, dribbling of saliva, and bleeding from the mouth, are also generally present.

Treatment.—The teeth should be brought together, thus allowing the superior maxillary bone (upper jaw) to act as a splint. A four-tailed bandage (Fig. 43) should then be applied to retain the parts in this position.

FRACTURE OF THE SPINAL COLUMN should be suspected when, following an injury to the back, there appears more or less paralysis *below* the point of injury (paraplegia), as the result of pressure upon the spinal cord. The ordinary symptoms of fracture are not usually present, *nor should they be sought for*, as efforts made to elicit crepitus, etc., might increase the injury to the spinal cord. Paraplegia may also be caused by a dislocation of the vertebræ.

Treatment.—The patient should be allowed to assume the position (lying down) which is most agreeable to him, provided he does not lie *face downward*. An ice-bag, or some form of cold, may be applied to the seat of injury, and perfect rest enjoined. If necessary, the patient can be moved on a stretcher which is so prepared that undue movement of the spine is prevented (see TRANSPORTATION). The after-treatment, consisting of efforts made to adjust the fragments of bone and to prevent subsequent inflammation and injury to the spinal cord, and relieving the paralyzed bladder of urine, should be attended to *only* by a surgeon.

FRACTURES OF THE RIBS are caused by direct or indirect violence, as the result of a fall or a blow, or being subjected to severe pressure, as in a crowd, or by muscular contraction. Muscular action during a paroxysm of coughing or sneezing, has been known to produce a broken rib. The seat of fracture is usually between the third and eighth ribs, and about on a line downward from the axillary space. The floating ribs—eleventh and twelfth—are rarely broken, owing to their single attachment (vertebral column), which allows sufficient freedom whereby to escape injury.

Fractured ribs are usually difficult to recognize, in a large number of cases the injury being mistaken for contusions.

Embarrassed and shallow breathing, associated with a sharp and lancinating or "stabbing" pain at the injured part, or a "stitch in the side," is almost always complained of. Crepitus may sometimes be detected by placing the hand or the ear against the injured side, and then having the patient take a long breath or cough. It is uncommon to find any external evidence of fracture, and great care should be observed in examining so as not to inflict a greater injury by the manipulation.

A broken rib may be followed by serious consequences, as an injury to the lung, which would be demonstrated by shock, spitting of blood, and, in some cases, a crackling sensation when the hand is carried over the skin at the seat of pain, due to the presence of air under the skin (emphysema).

Treatment.—The treatment consists in limiting the action of the affected side or sides as much as possible. This indication is met by the application of bandages or adhesive plaster. A triangular bandage, folded in the form of a cravat, and bound snugly around the chest, would answer for a temporary dressing; a flannel or muslin bandage, about three inches wide, and made to encircle the chest a number of times, would be still better. At the present time, however, the most effective means of treating a fractured rib is by the use of adhesive plaster, applied in the following manner: Strips of plaster, one and a half or two inches wide and about fourteen to twenty inches long, (for the adult) or sufficient to extend somewhat more than half-way around the chest, are made ready. The strips are heated and then applied firmly around the chest (from above downward), following, as nearly as possible, the course of the rib, each strip overlapping the lower third of the preceding one. The space covered by the strips should be about eight inches in width. If, for example, the fracture is on the right side, the adhesive plaster should be first applied at about three inches to the *left* of the spinal column, and the other end should be carried around the *right* side of the chest to about the same distance to the *left* of the breast-bone, thus restraining the action of the injured side without materially affecting the opposite one. In the use of bandages or the strips of plaster, as already described, it is very important that they (particularly the plaster strips) should be applied *at the end of expiration*, as at this time the chest is diminished in size and the broken fragments are brought closer together.

The above treatment may also be used in severe contusions of the chest.

FRACTURE OF THE CLAVICLE.—The collar-bone is more frequently broken than any other bone in the body, and is usually the result of indirect violence, as falling upon the shoulder. The recoil of a gun may cause direct fracture. It will be remembered that the collar-bones hold the shoulders *upward, backward,* and *outward;* consequently, the deformity following this injury would be dropping of the shoulder *downward, forward,* and *inward,* with some change in the outline at the site of the fracture.

Treatment.—The patient may be laid down, with a pillow between the shoulders, or, if he is to be removed, the following temporary support may be used: A soft pad is placed well up in the axilla or arm-pit, the forearm being laid against the chest and the shoulder raised by pressing the elbow upward, and held in this manner by an assistant, or by the patient himself, until supported by the bandages to be now described. A triangular or Esmarch bandage, folded in the form of a cravat, or a long strip of cloth, should be placed under the elbow, and one end carried upward, across the chest, and the other across the back, and the two ends fastened over the opposite shoulder; a similar bandage is then placed against the outside

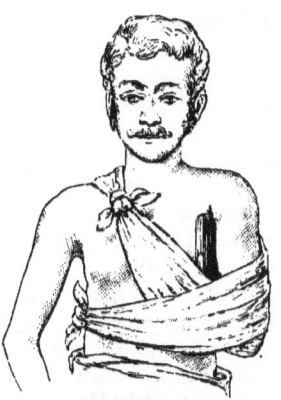

FIG. 75.—Temporary dressing for fracture of clavicle.

of the arm, on the injured side ; the ends of the second bandage are carried across the chest and fastened at the opposite side. The dressing fulfills, to a great extent, the indications for treatment, viz., carrying the shoulder upward, backward and outward (Fig. 75).

FRACTURE OF THE SCAPULA or shoulder-blade is of rare occurrence, and the result of direct violence, as a heavy wagon-wheel passing over the part.

The signs and symptoms are usually obscure, the bone being so thickly covered with muscular tissue that it does not exhibit much deformity.

Treatment.—In a suspected fracture, the arm of the corresponding side should be placed in a sling and kept quiet by the side of the chest.

FRACTURE OF THE HUMERUS or arm-bone may be the result of direct or indirect violence or of muscular contraction. The signs and symptoms are usually well marked, and there is considerable shortening. If the seat of fracture is at the surgical neck, a bony prominence may be found in the axillary space or arm-pit.

Treatment.—When reduction is to be made, the forearm and elbow should be pulled downward (extension), the shoulder acting as the counter-extension. If the fracture is about the middle of the bone, an internal and external splint should be applied. Great care must be taken least the internal one should be carried into the armpit and made to press upon important blood-vessels and nerves. If the fracture is in the upper part of the bone, through the surgical neck, a soft pad may be placed in the arm-pit and held in place by a figure-of-8 bandage, or spica, around the shoulder. The hand and wrist only should be suspended in a sling, thus allowing the elbow to drop, thereby diminishing the tendency to shortening of the arm.

FRACTURE OF THE FOREARM generally occurs at the lower end of the radius, about two inches or less from the joint, and is known as Colles' or "silver-fork" fracture.

It receives the latter name from the peculiar appearance about the wrist, which has a fancied resemblance to a silver fork. Colles' fracture is almost as frequent as fracture of the clavicle, and is generally impacted.

Treatment.—The arm should be placed at right angles, the thumb pointing toward the chin—midway between pronation and supination. An internal splint, extending from the upper part of the forearm to the wrist, and an external splint extending to the base of the fingers, should be applied. Care should be taken that the internal splint does not press upon the vessel at the bend of the elbow.

Reduction may be accomplished as follows: Should the fracture be on the right side, for instance, the attendant should grasp the right hand of the patient with the corresponding one of his own, and extension be carefully made; counter-extension being performed with the left hand, which grasps the forearm above the seat of fracture.

FRACTURE OF THE MIDDLE OF THE FOREARM may occur as the result of direct violence. One or both bones may be broken. When the latter occurs, the appearance is very characteristic. However, when either the radius or ulna is broken, the companion-bone acts as a splint, and the deformity and other signs are not so marked. In fracture of the *shaft* of the radius, which is uncommon, the usual symptoms of fracture are noted, and, in addition, the loss to a greater or less degree of pronation and supination, or turning the hand inward and outward. Fracture of the ulna sometimes follows an attempt to ward off a blow, and is not uncommon among pugilists.

Treatment.—Fracture at the middle of the forearm should be treated in the manner already described for Colles' fracture—internal and external splints applied while the arm is bent at an angle, with the thumb pointing to the chin, the injured limb then being supported by a sling.

FRACTURE OF THE METACARPAL BONES.—Indirect violence, as a fall upon the hand, or striking a blow with the fist, is a common cause of fracture at this situation. A swelling on the back or dorsum of the hand usually follows, and the knuckle corresponding to the broken metacarpal bone is sunken, and appears to be effaced.

Treatment.—A roller-bandage, wad of cotton, oakum, or other similar material, or a potato, lemon, tennis ball, etc., should be placed in the palm of the hand and retained there by a bandage.

FRACTURES OF THE PHALANGES are generally detected without much difficulty, the common signs of fracture being well marked.

Treatment.—After reduction has been accomplished, splints should be applied to the palmar and dorsal sides of the broken finger. A piece of a cigar-box may be used for this purpose, although a piece of tin or sheet-zinc, entirely covered with adhesive plaster, would be preferable; or, after the fracture is reduced, the finger may be bound to a companion finger, or covered with a narrow roller-bandage, and stiffened (after being applied) with flour and white of an egg, which makes a very good temporary dressing.

FRACTURE OF THE FEMUR is one of the common fractures of the body. The great size of the bone—it being the largest in the skeleton—and the fact that more or less limping due to the shortening of the affected limb may follow this injury (particularly in the adult), makes the treatment a matter of great importance to the surgeon. The femur may be broken either at the neck, extremities, or shaft, the most frequent situation being about the middle of the bone. The fracture is more commonly caused by indirect violence, as falling, etc., and is oblique (in adults), which principally accounts for the shortening that follows.

The signs and symptoms are usually well marked. The foot and leg are turned outward, particularly if the shaft of the bone is broken. Fracture of the neck of the femur occurs in old people, and is in a great measure due to the composition of bone at this period of life, which is more dense than at any other time, and also to the change in

FRACTURES. 149

the angle of the neck of the bone, and may in the aged follow a most trivial cause, as tripping, etc.

Treatment.—A long external splint, extending from about four inches below the axillary space or arm-pit to a short distance below the foot, is sufficient for a temporary dressing. If a gun is used as a splint, the stock should be placed beneath the arm-pit, and the barrel (turned downward) laid against the leg (Fig. 76). A fence-board, which is usually about six inches wide, makes an excellent splint. Before applying the splint, the thigh should be surrounded by a coat properly folded, or shawl, etc. A bandage should then be carried around the waist, two around the thigh (one above and one below the seat of fracture), one around the leg and foot each—the bandages being tied on the outside of the splint. If no better support can be devised, the affected limb may be bandaged to the one of the opposite side.

If a fracture of the femur can not be attended to by a surgeon within a short time, temporary extension and counter-extension can be made. The patient, for example, can be laid upon a bed, the foot of which is raised six or eight inches, thus carrying the body of the patient away from the seat of fracture (counter-extension).

FIG. 76.—Gun used as a temporary splint for fracture of the femur.

A large wad of cotton, or a fold of flannel or other soft material, should be wrapped around the ankle and instep to prevent the bandage about to be described from injuring the part. The next step consists in arranging some means by which temporary extension can be made. This indication is very effectively met by the use of Gerdy's "extension-knot," which not only holds the foot in a firm grasp, but prevents the undue constriction that would ensue if an ordinary knot were employed for this purpose.

Gerdy's knot can be made out of a strip of muslin five or six feet long and about six inches wide. This should be folded into a cravat (two inches in width).

The center of the cravat is placed upon the ankle-joint behind, the ends being brought forward and crossed over the instep. They are then continued downward under the sole of the foot, recrossed and carried upward on the sides to the malleoli (the bony prominences at the inner and outer sides of the ankle-joint), and then *under* the first turn around the ankle and downward, and tied below the foot in the form of a loop, in order that the weight shortly to be described may be attached (Fig. 77). A pail, bag, or some other receptacle should now be procured and in it placed stones or sand, or weight in some form, amounting to from ten to fifteen pounds, and so arranged that it can be readily attached, at the proper time, to the loop connected with the patient's foot. Extension and counter-extension should now be made by the attendants, as follows: While one holds the body in position, another grasps the leg and foot on the affected side, and makes steady extension; a third attendant should place his hands over the seat of fracture to support and protect it.

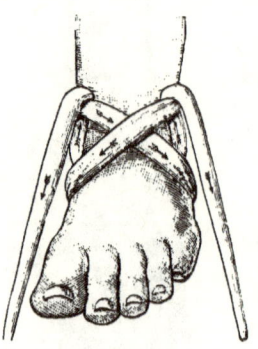

Fig. 77.—Gerdy's extension-knot.

When the proper length has been regained, or nearly so, and while the limb is held in this position, the weight already prepared should be carefully and slowly attached. The weight should hang suspended over the foot-board of the bed, if the latter is low enough (on a level with the affected extremity), or over a board placed upright. *After* extension and counter-extension are made, and the weight is attached, the splints should be applied—*not before*. If the attached weight causes pain, it should be diminished.

FRACTURE OF THE LEG.—Fractures of this portion of the body usually affect both bones, although either the tibia or fibula may alone be fractured, the fibula probably oftener than the tibia. The fibula is generally broken near its lower extremity, and is known as Pott's fracture. More or less injury to the ankle-joints accompanies the latter. In fracture of both bones, which commonly occurs about the middle, the signs are generally

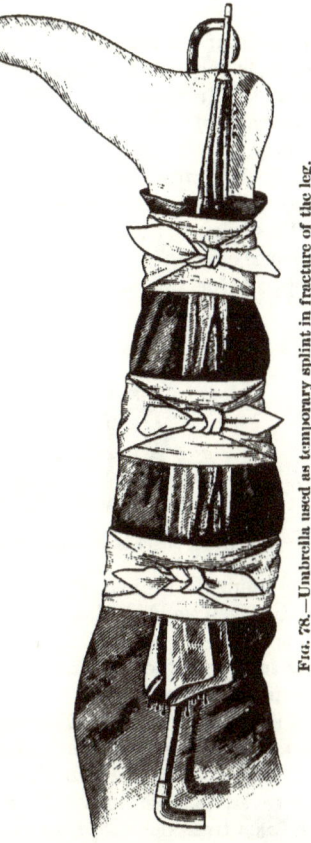

Fig. 78.—Umbrella used as temporary splint in fracture of the leg.

well marked, more so than when either the tibia or fibula only is broken. Pott's fracture is attended with eversion or turning outward of the foot, producing a characteristic deformity.

Fracture of the tibia, as the result of its superficial situation, is very often compound; the wound communicating with the fracture being frequently caused by efforts on the part of the patient to walk immediately after the injury.

Treatment.—The limb should be handled very carefully, and an internal and external splint applied. A pillow placed under the leg, folded over the sides, and properly retained, is particularly adapted as a temporary support for this fracture (Fig. 79). If the lower end of

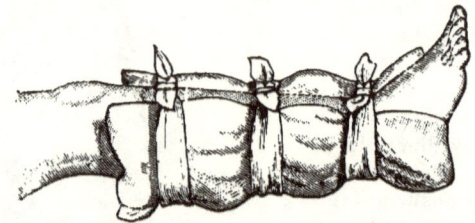

Fig. 79.—A pillow used as a temporary splint in fracture of the leg.

the fibula is broken, and the foot is turned outward, a splint, well padded, should be placed along the *inner* surface of the leg, extending from above the knee to beyond the foot, and the leg and foot bound to it, thereby overcoming the tendency to eversion.

Fracture of the patella may be the result of direct violence or muscular action. The more common signs of the injuries are inability to straighten the leg; the patient, however, is able to walk backward. An examination shows a transverse separation of the knee-cap, with an interval varying in width between the pieces. The knee is swollen, tense, and painful. At the moment of the injury the patient often hears a sudden snap.

As a temporary measure, a long posterior splint may be applied, carefully bound above and below the knee. The skill of the surgeon is directed toward keeping the pieces in close contact until union takes place.

THE TREATMENT OF FRACTURES OF THE METATARSAL BONES AND PHALANGES OF THE TOES consists in binding the affected toe to the one next to it, or by supporting the fragments with compresses or light splints. When the foot is seriously injured by crushing, amputation is often inevitable.

DISLOCATIONS.

A DISLOCATION or luxation is a forcible displacement of one articular (joint) surface of a bone from another, and may be the result of direct or indirect violence, or of muscular contraction. More or less rupture of the ligaments always takes place. The chief signs of a dislocation are deformity and loss of function of the joint. The mobility of the part is greatly diminished, while in fractures there is increased mobility. In dislocation the deformity is at the joint, while in fracture it is usually about the shaft.

The reduction of a dislocation requires considerable technical skill, and should be performed by a surgeon. Exceptions to this rule, however, may be made in dislocation of the *shoulder, lower jaw,* and *fingers.*

DISLOCATIONS OF THE HUMERUS usually take place downward below the coracoid process (see SCAPULA). The indications for reduction are, to disengage the head of the humerus from its abnormal position by extension, counter-extension, and fulcrumage. This may usually be accomplished in the following way : A firmly compressed ball of cotton or similar material should be placed in the axilla or arm-pit ; the attendant should then remove the shoe from his foot nearest the affected side of the patient (facing the latter), and press the heel upward against the ball already in the arm-pit ; he should also grasp the patient's hand and arm, and pull downward, thus making extension and counter-extension. In this manner the head of the humerus is disengaged, and, if the foot of the attendant can be turned outward and the arm of the patient brought toward the chest *during extension and counter-extension*, the bone will generally slip back into its proper position. After the reduction the arm should be bandaged to the chest, to prevent redislocation. It should not be forgotten that the axillary space contains many important blood-vessels and nerves, and too forcible manipulation may be followed by very serious consequences.

DISLOCATION OF THE LOWER JAW (inferior maxillary bone) is comparatively rare. It may be the result of a blow, but usually follows the act of gaping, laughing, or vomiting. The deformity is striking: the jaw is protruded and remains open, the patient being unable to bring the teeth together. The articular surfaces of the bone in this injury being carried *forward*, and somewhat upward, the reduction should be accomplished by depressing the articular portion (condyles) of the lower jaw, and forcing them backward. This may be performed in the following manner: The patient should be placed in a chair, with the operator standing before him, having his thumbs wrapped in a handkerchief, or some similar material, to guard against injury during the sudden closure of the patient's jaws at the moment the dislocation is reduced. The attendant then places a thumb upon each posterior molar tooth of the inferior maxillary bone of the dislocated jaw, and presses firmly downward, and with his fingers he tilts the chin upward. While the bone is being carried downward in this manner, a backward pressure with the thumbs should then be added which helps to carry the articular surfaces to their proper position. The jaw should then be held in place by a four-tailed bandage (Fig. 43).

DISLOCATION OF THE PHALANGES may be reduced by bending the dislocated bone further back, at the same time making extension and counter-extension, then suddenly flexing the joint. The subsequent adjustment of a dorsal and palmar splint is necessary.

SPRAINS.

A SPRAIN is a wrenching or twisting of a joint, associated with considerable stretching, and even tearing, of the tendons and ligaments of the affected part. Sprains usually occur at the ankle or wrist. Pronounced swelling and pain rapidly ensue, and are characteristic of the injury; though it is often difficult to differentiate between a sprain, dislocation, or fracture. The proper relation of the ends of the bone composing the joint, and the absence of the principal signs of dislocation or fracture, indicate that a sprain exists. A sprain is always troublesome, and may be followed by serious results, a common sequel being anchylosis, or stiffness of the joint.

TREATMENT.—When the patient is seen soon after the injury, the part should be elevated, and a cold application made, preferably the rubber bag (manufactured for this purpose), or a pig's bladder partly filled with cracked ice. Should neither of these be obtainable, the cracked ice may be placed in a handkerchief or towel, and, if possible, the dressing enveloped with oiled silk or rubber cloth. If ice can not be procured, cloths wrung out in cold water should be substituted. A roller-bandage, carefully applied from the extremity upward, is also valuable in preventing the continuation of the swelling. The ice or cold-water applications can be applied *over* the bandage; it must, however, be borne in mind that a bandage that has been wet will shrink, and may make too much pressure, or even cause strangulation of the part. After the acute symptoms have subsided and the heat and swelling have diminished, the cold applications should be discontinued and gentle friction substituted; or the joint may be rubbed with a stimulating lotion, as soap-liniment, alcohol, or salt water. Massage is also a valuable remedy. In severe sprains affecting the larger joints these parts should be kept quiet for two or three weeks, and motion then be gradually performed.

CHAPTER XIII.

BURNS, SCALDS, AND FROST-BITE.

BURNS AND SCALDS.

BURNS are caused by the action upon the tissues of some form of dry heat, or by a chemical agent. For practical purposes burns are divided into three degrees: (1) Simple redness of the skin; (2) vesication or the formation of blisters; (3) more or less destruction or charring of the skin and deeper structures. Burns of the first and second degrees are usually unattended by serious consequences, but when they involve one half or more of the surface of the body a fatal result usually follows. Burns of the third degree are dangerous according to their situation, extent of injury, and complication. Burns of the thoracic and abdominal walls are often attended with great danger. Death following burns is the result either of shock (during the first twenty-four hours), internal inflammation, ulceration and hæmorrhage, blood-poisoning, tetanus, or exhaustion.

The temperature of the body falls immediately after a severe burn; this is temporary, however, and is soon followed by more or less fever.

In burns of the third degree, which heal by granulation, the contraction of the scar and subsequent deformity of the part are particularly marked.

TREATMENT.—The treatment of burns is divided into local and constitutional. The *local* treatment depends upon the degree to which it belongs.

In BURNS OF THE FIRST DEGREE, remedies which are

soothing and protective against atmospheric air and cold should be applied. Among those which generally can be procured at once are bicarbonate of soda (common baking-soda, *not washing-soda*), starch, flour, chalk, magnesia, or charcoal. One of these may be thickly dusted over the burned surface. Vaseline, cosmoline, lanoline (made from wool-grease), olive, linseed, or castor oil; also lard and butter, provided they are not salted nor rancid; white lead paint, or lime-water, whitewash, or even ink, mucilage, or molasses may be employed, although inferior to some others.

The agents enjoying the highest reputation in the different hospitals, at the present time, are those containing carbolic acid (about five per cent), as carbolated cosmoline, vaseline, oil, lanoline, etc. The carbolized applications are not only protective and antiseptic, but the carbolic acid which they contain produces anæsthesia of the affected surface; that is, it diminishes the sensibility of the skin. "Carron-oil," composed of equal parts of linseed or olive oil and lime-water, is regarded as a very valuable remedy. The cosmoline, vaseline, carron-oil, etc., should be spread on a piece of lint, soft linen, or muslin, and laid on the burn. The part should then be enveloped in cotton batting, or two or three folds of flannel, or some other soft material which will properly protect the part ; the dressing should then be retained by a bandage. During the examination of severe burns of any degree, and also while the application of dressings is being made, great caution should be observed that the part is not unduly exposed to the air ; death has followed the careless exposure of a burn of this character. Such danger can be avoided by examining or dressing one part of the burn at a time.

In BURNS OF THE SECOND DEGREE the blisters require special treatment. If attached to the burn, the clothing should never be forcibly removed, but carefully cut off with scissors as close to the burn as possible. The small pieces adhering to the skin may be washed away with warm water, or softened with oil, and detached later. If the blisters are large, they should be pricked at their lowest part, and the contents allowed to escape, or should be absorbed with clean blotting-paper. The oily substances recommended for burns of the first degree may then be used, or the part carefully washed with an antiseptic solution, and dry antiseptic gauze applied. The dressings should not be disturbed oftener than every two days unless they become moist or offensive.

In BURNS OF THE THIRD DEGREE, where there is destruction of the tissues and more or less sloughing, antiseptic applications (carbolized vaseline, oil, etc.) are particularly indicated; offensive discharges should not be allowed to accumulate, but must be removed by warm antiseptic solutions, and fresh dressings applied. The removal of the slough or dead tissue may be hastened by the use of antiseptic poultices. (See POULTICES.)

In burns caused by acids (generally sulphuric, nitric, and muriatic), water should not be applied, for, when combined with an acid, an elevation of temperature of the mixed fluids immediately follows. The proper remedy would be the application of an alkaline powder, bicarbonate of soda, magnesia, chalk, or lime; the latter may be scraped from the walls. These agents neutralize the acid, and should be left on the surface but a few moments and then washed off.

If an acid is splashed into the eye, lime-water or a solution of soda or magnesia should be applied at once, also a few drops (three or four) of a four-per-cent solution of cocaine; the latter relieves the excessive pain. Burns of the mouth and throat from the same cause (acids) are also to be treated by the free use of the alkalies already mentioned. (See POISONS.)

Burns produced by caustic alkalies (caustic soda, potash, lime, ammonia, also quicklime and lye) should be treated by the application of acid solutions, as diluted vinegar, lemon-juice, hard cider, etc. Nitric, muriatic, sulphuric, and acetic acid may be used, but only when *well diluted*, and with great circumspection.

After the acid or alkali causing the burn has been neutralized, the oily substances already referred to should be applied as in ordinary burns.

CATCHING FIRE.—Some of the severest forms of burns follow the catching fire of some portion of the wearing-apparel. A person in this condition should at once be enveloped in a blanket, coat, mat, piece of carpet, or whatever may be at hand to extinguish the flames, or she (the victim is usually a woman) should be rolled over and over on the floor to smother the flames. Water must be freely used and the clothing carefully examined to ascertain if the fire has been entirely extinguished. The burned surfaces should then receive the proper attention, according to their degree and intensity.

Constitutional Treatment of Burns.—Shock, which is always present in severe burns, requires the administration of stimulants. Pain is more constant and intense in burns than in any other form of injury, and requires sedatives, which should be administered by the medical attendant. Later on, the appearances of inflammation and other complications are to be carefully watched for.

SCALDS are injuries produced by the application of moist heat, as boiling water, steam, etc. Children commonly suffer from these injuries as the result of pulling over kettles containing hot coffee, tea, or water. Scalds should be treated as burns of the first and second degrees.

FROST-BITE.

Prolonged exposure of the body to a very low temperature results in a general or local loss of vitality. While the air is still, or snow is falling, it is favorable to the one exposed. Snow is a bad conductor of heat, and offers considerable protection; but when the wind is blowing, the warm air close to the surface of the body is rapidly removed, and the destructive effect of the cold is considerably increased. An exposure of one or two hours, inadequately clad, to intense cold may be followed by a fatal result. Those who are thus unfortunately situated are soon overcome by an irresistible sense of drowsiness and desire to sleep; to yield to the inclination is usually fatal. This disposition to stupor is due to the great diminution in the blood-supply to the surface, and consequent congestion of internal organs. In this condition the brain is unable to properly perform its function and the drowsiness follows.

In those who are frozen, the limbs become stiff and the skin white; the latter is preceded by a blue or purplish tint, which may still be apparent at the tips of the nose, toes, and fingers. This change in color denotes that the circulation and nourishment of the surface of the body are profoundly interfered with, and is commonly followed by excessive reaction and inflammation or gangrene.

TREATMENT.— A person who is frozen should never be taken to a warm room, or have warmth applied to the body. An abrupt change in temperature would be almost necessarily fatal. The temperature must be gradually raised. Consequently, the patient should be carried to a cool apartment, the clothing removed, and the body rubbed with snow or cold water. After a short time, particularly if consciousness returns and the limbs lose their rigidity, the use of a piece of flannel, or, still better, the hand, can be substituted for the cold application. The continuous rubbing may now be discontinued, and occasional friction resorted to. These measures must be pursued very gently, as rough manipulation might destroy the skin. It may be necessary to resort to artificial respiration in extreme cases.

Stimulants carefully administered are indicated if the patient can swallow; until this is possible, they should be given by the rectum, or ammonia or smelling-salts by inhalation ; nourishment in the form of beef-tea or milk may be given as soon as the patient can take it. The surface of the body should be carefully protected, but not subjected to heat, as some time must elapse before the circulation of the affected part regains its equilibrium. If portions of the surface subsequently become dark bluish or mottled, gangrene has commenced.

CHAPTER XIV.

UNCONSCIOUSNESS, SHOCK, AND SYNCOPE.

UNCONSCIOUSNESS is associated with a number of different affections, and in order to intelligently treat this condition it is necessary that one should be familiar with the different causes producing it; among which are apoplexy, opium-poisoning, alcoholism, syncope, concussion and compression of the brain, and epilepsy. This is very important, as the remedies suitable for one would be unfit for another: for instance, the unconsciousness due to syncope demands stimulants, which, if administered in apoplexy, might cause or hasten a fatal result. Ignorance of this subject has often been responsible for the humiliating and serious mistake of regarding a case of apoplexy as intoxication.

As the result of complications, it sometimes happens that the cause of the unconsciousness may not for the moment be apparent; however, a careful examination for signs indicating the different affections giving rise to this condition will generally clear up existing doubts. During this period of uncertainty the patient should be placed in the recumbent position, with the head slightly elevated or depressed, according to the appearance of the face, whether congested or pale. The clothing about the neck and body should be loosened, and fresh air freely supplied. Stimulants should be used with caution, and only when positively indicated. Artificial respiration may be resorted to in extreme cases.

SHOCK—COLLAPSE.

Shock is a condition in which there is a more or less diminished energy of the heart and circulation, and is the result of a severe impression made upon the nervous system, produced by either a physical injury, or a mental emotion. The majority of cases met with are the result of extensive burns or other grave injuries, particularly those produced by gunshot wounds and railway accidents, which are generally associated with great laceration and crushing of the tissues, and mental excitement. Severe cases of shock may be produced by fright alone. Shock may be of a very mild character, as the result of a trifling injury or fright, the symptoms being hardly noticeable, of short duration, and demanding no treatment; or, it may assume a form which is rapidly fatal.

The symptoms of shock depend upon the severity of the cause; in some, where the injury is slight, they may be hardly apparent, or, only a pale face and a weak and rapid pulse, a slight nausea, and a general sense of prostration may be evinced. The form that fully illustrates a case of severe shock would be that following a serious railway injury, and can hardly be mistaken. There is no other condition which so closely resembles death. The extreme pallor and coldness of the skin are startling; the surface of the body is covered with moisture; large beads of sweat cover the forehead; the pulse at the wrist may be lost, or if perceptible, is weak, rapid, and irregular; the features are shriveled, particularly about the nose, which appears pinched; the eyes are lusterless, sunken deeply in the sockets, and turned upward, the pupils being generally dilated; the fingers and nails are of a bluish color. The patient is conscious, but dazed and flighty, can not realize his condition, and apparently only appreciates loud and repeated questions; articulation is difficult, although there is no paralysis present. The sensibility to pain may be so blunted that an operation can be performed without the cognizance of the patient. As a result of the depression of the circulation, the heat of the body is diminished, the temperature in some cases being below the normal register. The respirations are sighing in character and irregular, and the patient is restless. These symptoms may continue for a few minutes or a number of hours, and often end in death.

In cases which end favorably, there appear, usually within an hour or so, symptoms denoting an increase in the strength of the heart and circulation ; this change is known as reaction, and generally indicates, so far as the shock is concerned, a happy termination. When reaction occurs, the color and warmth gradually return to the skin, the eyes are brighter and the mind clearer; the pulse becomes stronger, and the symptoms indicate an approach to the normal condition. Vomiting is regarded as a favorable symptom, and generally denotes reaction.

Reaction, however, does not always insure the safety of the patient, as it may be interrupted suddenly by hæmorrhage or failure of the heart and death; or, in injuries about the head, the reaction may be so intense as to go beyond the normal activity of the circulation, and produce serious cerebral diseases, such as congestion or inflammation of the brain; for this reason the patient should be carefully watched, and excessive reaction obviated by position, quiet, cold applications to the head, and warmth to the extremities.

Cases of shock from severe injuries, associated with complete unconsciousness, almost always prove fatal.

TREATMENT.—It will only be necessary to refer to the condition indicated by shock to clearly understand the means necessary to bring about reaction; consequently, whatever can safely be used to stimulate the heart and circulation constitutes the proper treatment. As an illustration, we will suppose that a man has been injured by a railway accident and is found in a condition of shock. Whoever is to attend to the case should at once loosen the clothing, and make a rapid examination to ascertain whether severe hæmorrhage exists (which fortunately is not common in shock, owing to the feebleness of the circulation), or whether some other symptom may be present demanding immediate attention, which being promptly alleviated, the patient should be carefully conveyed to a place near by, more suitable for the subsequent treatment. While being removed the head should be as low as or somewhat lower than the body, or the extremities may be slightly elevated, so as to favor the flow of blood toward the brain. If possible, four persons should be selected to carry the patient, one for each extremity and the contiguous portions of the body. If one or more of the bones of an extremity have been broken, a temporary splint ought to be applied to prevent any movement of the fragments of bone at the seat of fracture during transportation, and should be attended to as rapidly as possible, and before the man is lifted from the ground. Arriving at the place selected, the clothing of the patient should be removed carefully, or cut away, if necessary, rather than have too much delay. He should then be placed in a warm bed, his head being still kept low.

The treatment now consists in applying warmth to the surface of the body, and internal stimulation. The first indication can be met by surrounding the patient with bottles containing hot water, placing them about the arms and legs, inside the thighs, and under the arm-pits and about the body, but not about the head, as the heat might favor a subsequent congestion when reaction occurs; hot bricks,stones, plates, etc., will also answer the purpose, dry heat being the essential element. It must be remembered that the sensibility of the patient is blunted, and that the bottles or bricks may be so hot as to burn the skin and not inflict any pain. A hot plate, enveloped in a towel, may be placed over the region of the heart, and, when unusual vomiting occurs, mustard plasters applied over the stomach. Friction is a method of some value in exciting the circulation, and should be resorted to when heat can not be supplied as indicated above. The second indication —the use of internal stimulation—is of course governed by the condition of the patient. If able to swallow, he should be given about two teaspoonfuls of whisky or brandy, with a small amount of hot water, or, still better, hot milk; this may be repeated every ten or fifteen minutes, until four or five doses have been taken, or reaction becomes apparent. When the latter occurs, the stimulant should be diminished or discontinued. There are cases where a greater amount of liquor may be required, but it should be given with caution, as too much may produce undue vomiting and favor excessive reaction. If a number of large doses of a stimulant have been given with no effect, it is evident that the stomach is unable to absorb the fluid, and its further administration by the mouth would be useless. A large number of persons suffering from shock are unable to swallow, as the result of extreme prostration ; this should be carefully ascertained before giving anything by the mouth, otherwise strangulation might ensue, by the fluid entering the larynx. In such cases a tablespoonful or more of liquor in half a cupful of warm milk or water can be introduced into the rectum (lower bowel) by a syringe, and may be repeated about every fifteen or twenty minutes ; also injections of hot water, a pint at a time.

The value of the liquors depends upon the percentage of alcohol they contain, which in brandy and whisky is about forty to fifty per cent, and consequently more than in wines, which contain but from fifteen to thirty-five per cent. One half the amount of diluted alcohol would therefore be a substitute for the liquors. Beer and ale, containing only two or three per cent of alcohol, would be worse than useless, simply filling up the stomach, with no satisfactory results. A small amount of spirits of ammonia or ether, or about four or five drops of nitrite of amyl, on the hand or a handkerchief, and placed near the patient's nose, has a decided stimulant effect. Warm turpentine rubbed up and down the spine can be used with benefit, if it does not interfere with the treatment already begun.

The usual and most effective method of introducing stimulants into the system, now employed by physicians, is by the hypodermic method, which consists of injecting under the skin, with the hypodermic syringe, stimulants—brandy, whisky, or ammonia-water; the amount so injected being generally from twenty drops to a teaspoonful of the undiluted stimulant, and is repeated as often as is indicated by the condition of the patient.

Artificial respiration may be resorted to in desperate cases, but it is of doubtful value. When reaction has taken place, the stimulants are gradually diminished, and some form of nutriment may, if required, be administered; it must be given in small doses, and of such a character as to be readily absorbed. Warm beef-tea or milk, or an occasional sip of kumyss, will be sufficient for the time being.

SYNCOPE—FAINTING.

Syncope is a condition of suspended animation associated with a great diminution of blood in the brain and unconsciousness, and caused by sudden enfeeblement of the heart's action.

Syncope may be the result of disease of the heart, hæmorrhage, pain, excessive emotion, as grief, fear, and joy; tight lacing, certain drugs, indigestion, hunger, exhaustion, and the hot and vitiated air of public assemblies, are also exciting causes of this condition. Before the stage of unconsciousness, the person affected experiences a weak and sinking feeling, associated with dizziness, dimness of vision, roaring in the ears. The face and extremities become pale, cold, and clammy.

The stage of unconsciousness may last a few seconds or a number of hours; however, it usually extends over a period of but a few minutes, during which time the pulse is either very weak or lost, and the respirations are shallow and sometimes apparently cease.

Fainting is of common occurrence and not usually fatal.

Treatment.—The treatment is very simple, and consists in stimulating the heart and circulation, and increasing the amount of blood in the brain. The patient should be laid down, with the head somewhat lower than the body, to favor the flow of blood toward the brain. Fresh air is one of the essential elements in the treatment of fainting. All tight clothing, as collars, corsets, skirts, or trousers, should be at once loosened, and stimulation in some form begun. Smelling-salts, ammonia, and other stimulants may be poured on a handkerchief or the palm of the hand and given by inhalation, care being taken that none is dropped into the eyes, and that the remedy is not held too close to the nose or mouth, otherwise violent irritation of the air-passages may follow. The head and face may be bathed with an alcoholic solution, as camphor, bay-rum, cologne, or whisky, or friction can be applied to the limbs and over the heart. No attempt should be made to administer stimulants by the mouth unless the patient is able to swallow.

The means just described are usually sufficient to bring about reaction. In some cases, however, it will be necessary to resort to more active measures, such as the introduction of stimulants into the rectum or lower bowel, in the manner described in shock, or the inhalation of nitrite of amyl —about five drops on the palm of the hand or on a handkerchief—held to the nose of the patient, the application of mustard over the heart, or it may be necessary to employ artificial respiration.

A person after recovery from an attack of syncope should be kept quiet until the action of the heart and circulation is properly strengthened.

Syncope and shock are similar, and may be the result of the same cause; however, in syncope the patient is *unconscious*, while in shock he is more or less conscious. Shock usually follows a severe injury, whereas fainting may follow a very trivial cause. Syncope differs from apoplexy in the following manner : In syncope, the face is pale, the pulse weak, and the respiration shallow, and the affection is generally of short duration; while in apoplexy the face is usually congested, the pulse about normal or slower, and the breathing is noisy and labored or snoring (stertorous), and paralysis exists on one side of the body.

CHAPTER XV.

*CONCUSSION AND COMPRESSION OF THE BRAIN—APOPLEXY
—INTOXICATION—EPILEPSY—HYSTERIA—HEAT-STROKE.*

CONCUSSION of the brain is a term applied to a shaking up or jarring of this organ. More or less contusion of the cerebral tissue probably occurs in severe cases. This form of injury is the result of a fall or blow upon the head, or the shock may be transmitted upward through the spinal column, as in jumping from a height, etc. The injured person may only be stunned or dizzy for a moment, with no loss of consciousness; the face is usually pale, and more or less weakness and trembling of the limbs are present. These transient symptoms are generally followed by a rapid restoration to the normal condition; although in apparently mild cases serious cerebral disturbance may subsequently occur—usually within a week. In severe cases there are partial stupor, feeble pulse, and contracted pupils; the surface of the body is cold, and restlessness and vomiting are usually present. As a rule, the breathing is natural. In grave cases the patient is in the condition of shock, which has already been described. There is one point connected with the shock following concussion of the brain which must be emphasized, viz., that the reaction is usually excessive, being more pronounced than in shock following injuries of other parts of the body, and often terminates in inflammation of the brain.

Treatment.—In mild cases the treatment consists in keeping the patient quiet, with a cooling application to the head. The more serious cases of concussion of the brain which assume the form of shock should receive the treatment appropriate to the latter affection, particular attention being paid to the extreme reaction that may occur. For this reason, internal stimulants should not be given. Warmth to the extremities, and cold applied to the head, are also indicated.

Rest and quietude must be insisted upon in all cases of concussion of the brain.

COMPRESSION OF THE BRAIN.

Compression of the brain commonly follows fracture of the skull, a portion of the broken bone being driven into the cerebral tissue and causing pressure. The lodgment of a bullet, or cerebral hæmorrhage, may also produce this condition. It is often difficult to differentiate between concussion and compression of the brain. However, in compression, the stupor is more profound, the pulse slow, the pupils dilated, and the breathing stertorous, similar to that in apoplexy. There is more or less paralysis present, and convulsions may occur.

Treatment.—Aside from keeping the patient quiet and applying cold in some form to the head, and preventing the administration of stimulants, very little can be done by an unprofessional person, and the case should be placed in the hands of a surgeon as early as possible. If a wound is present, it should be protected by an antiseptic compress.

APOPLEXY—STROKE OF PARALYSIS.

Apoplexy is a sudden loss of consciousness associated with paralysis, and generally due to the failure of a certain portion of the brain to perform its function, as the result of pressure following the escape of blood (hæmorrhage) from a diseased cerebral vessel. Embolism, which is very similar to apoplexy in its manifestations, is caused by the plugging up of a large artery within the skull, which is followed by a loss of vitality and function of a portion of the brain.

The paralysis is confined to one half of the body (hemiplegia). For example, if the hæmorrhage is on the left side of the brain, the muscles of the right arm, leg, and the corresponding side of the face are rendered powerless. The paralysis may sometimes affect the opposite side of the face. The paralysis occurring on the opposite side of the body from the seat of pressure is due to the crossing of the nerves already referred to (see NERVES).

The form of apoplexy caused by cerebral hæmorrhage affects persons advanced in years. A person may suffer from a number of attacks in quick succession, or at long intervals, or the first may prove fatal. The common belief that one dies during the third attack is unfounded.

An apoplectic attack is abrupt, and not generally preceded by any sign or symptom which indicates its approach. The subject usually falls to the ground as though struck down, although in some cases the attack is less sudden, and preceded by a sharp outcry.

If unconsciousness does not occur immediately, it follows in a very short time—within a few minutes. In this condition the patient can not be aroused, which is an important point in distinguishing this from some other affections. The face becomes reddened or congested, and the pupils are usually dilated—not uniformly, however; one may be dilated and the other in the normal condition or contracted. The respirations are slow, labored, and snoring (stertorous breathing), the cheeks being puffed out during expiration. The pulse, although generally slow, does not differ very much from that of health. Convulsions and vomiting may occur.

Apoplexy is often mistaken for intoxication—a very disagreeable and unfortunate error. It may also be mistaken for opium-poisoning and epilepsy.

It should be remembered that in intoxication, hemiplegia does not exist, and that the alcoholic odor of the breath is always present (this *may* sometimes be present in apoplexy); also that vomiting is common, and the subject can be more or less aroused by pinching, douching with cold water, etc.

In opium-poisoning there is an absence of hemiplegia, the respirations are very slow, sometimes being reduced to four or five during the minute. The face is pale and the pupils are uniformly contracted.

The absence of hemiplegia, the convulsive movements, and foaming at the mouth, and the comparative short duration of the attack, are usually sufficient to distinguish epilepsy from apoplexy.

If improvement does not follow an attack of apoplexy within ten or twelve hours, the case usually terminates fatally.

Treatment.—The object of treatment is to arrest the further escape of blood within the skull; consequently the head should be slightly elevated, and cracked ice or some other form of cold applied. The clothing about the neck and waist should be loosened. Internal stimulants, which tend to increase the hæmorrhage, must not be used. If possible, warmth should be applied to the extremities. The further treatment must be indicated by the medical attendant.

INTOXICATION.

Intoxication is a condition of such common occurrence that a lengthened description would be unnecessary. The important element in the consideration of this subject is to impress upon the reader the necessity of particularly distinguishing this condition from apoplexy, and has already been referred to in the article on the latter subject (see APOPLEXY). If any doubt exists, the patient should be laid upon his back, with the head slightly elevated, and all constriction about the neck and waist removed, and perfect quietude and rest maintained until the arrival of the medical attendant.

Treatment.—The treatment of intoxication consists usually of rest and quiet, Nature being generally able to effect the desired result. If vomiting does not take place, mustard may be given to produce emesis (vomiting). In some cases it may be necessary to apply warmth to the extremities and over the heart, also ammonia or nitrite of amyl by inhalation, and rectal stimulation (coffee, etc.).

EPILEPSY—EPILEPTIC FITS—FALLING-SICKNESS.

Epilepsy is an affection of the brain which at variable intervals gives rise to an attack characterized by convulsive movements and unconsciousness.

Subjects of this disease are usually warned of the approach of a paroxysm by certain sensations which immediately precede the convulsion; the premonitions, however, are generally of very short duration. The attacks usually occur singly, although one may follow another in rapid succession.

The person affected utters a sharp, piercing cry, and falls to the ground in an unconscious and helpless condition, and is thus frequently seriously injured about the face. This is followed immediately by more or less rigidity, and a few moments later the convulsive movements commence. The eyes are usually opened and turned upward, the pupils being dilated. The face becomes livid and congested, the jaws are brought together with considerable force, and the tongue is very commonly badly wounded by the teeth. There is considerable foaming or frothing at the mouth.

The paroxysm lasts but a few minutes, when consciousness gradually returns, the patient being in a stupid and drowsy condition for a short time afterward.

Malingerers and hysterical subjects who endeavor to simulate epilepsy do not bite their tongues, and usually select a convenient place in which to fall.

Treatment.—The treatment of an epileptic attack consists in laying the patient on his back, loosening the clothing, particularly about the neck, and preventing injury to the tongue by placing between the teeth a cork, piece of wood, or handkerchief-knot. The stupor that follows the convulsion requires no treatment, unless the patient is quite weak, when stimulants may be very cautiously given.

HYSTERIA—HYSTERICS—HYSTERICAL ATTACK.

Hysteria is a functional affection of the mind and nervous system, occurring in paroxysms, usually affecting females, and characterized by temporary loss of will-power and considerable emotional display on the part of the one affected, who laughs and sobs immoderately, without any regard to the surroundings.

Hysteria sometimes assumes a form which may be mistaken for syncope, apoplexy, or epilepsy, according to the manifestations of the patient.

Hysterical subjects should not be treated as malingerers; the perverted condition of the nervous system leads them to believe that the illness assumed is real. One unaccustomed to observing cases of hysteria may be easily misled, therefore the following distinctive characteristics should be remembered.

A case of hysteria may be distinguished from syncope in the following manner : In hysteria the extremities and face are warm instead of cold, and the pulse is normal; and when an effort is made to open the eyelids, it is met with considerable opposition, which does not occur in unconsciousness. In the hysteria that simulates epilepsy the tongue is not injured, and, if the patient falls, a suitable place is selected for this purpose, in order to avoid injury. There is no frothing at the mouth in hysteria, unless produced by soap or other agents held in the mouth.

Although hysteria may be associated with temporary paralysis and mistaken for apoplexy, it will be noticed that the facial muscles are not involved, and that the paralysis does not always affect one side of the body (hemiplegia), and there is no stertorous breathing and flapping of the cheeks, which are so prominent in apoplexy.

Treatment.—The best treatment for a case of hysteria is to let the patient alone. To sympathize with one in this condition simply prolongs the attack and can do no good. Firmness on the part of the attendant, with an intimation that the condition is fully understood, is generally all that is necessary. When considerable demonstration is made, it is a common practice to douche the patient with cold water. If this method is used, it should be confined to the face; pouring a large quantity of cold water indiscriminately over a weak and delicate woman is not always followed by the most desirable results.

HEAT-STROKE—SUNSTROKE—INSOLATION—SUN-FEVER— HEAT-APOPLEXY.

The term *heat-stroke* is now used to denote the serious manifestations which sometimes follow exposure to intense heat. This condition may be due to the direct rays of the sun, or to a high temperature caused by solar or artificial heat. Those who are employed in hot and close apartments, where the air is impure, as in mines and in the fire-rooms of steamers, are particularly liable to be affected by heat-stroke from artificial heat.

It is a fact worthy of notice that those who generally suffer from this malady are either addicted to the use of stimulants or are in a weak or debilitated condition.

Locality has considerable influence in causing sunstroke, the vitiated air of the city in its hot summer months being particularly favorable to it. The high degree of temperature which can ordinarily be borne by the body is mainly dependent upon the proper action of the skin; the large amount of perspiration being rapidly evaporated, cools the surface and indirectly the blood. Evaporation takes place more rapidly when the air is dry; consequently, heat-stroke is more frequent when the air is moist. The heavy woolen shirts worn by workmen in hot weather are cooler than those made of cotton, because the former increase the activity of the sweat-glands, and indirectly diminish the temperature of the body.

Heat-stroke is not always developed abruptly, nor does

it generally occur at the beginning of a heated term, but usually after the hot weather has been persistent for two or three days. There are certain symptoms which indicate the approach of this affection and should be recognized as a warning, viz.: an irritable and depressed condition; headache, congestion of the eyes, and flushing of the face, and also a dryness of the skin and sometimes nausea. Unless the person affected seeks relief, the symptoms just enumerated are followed by those which are more serious, as delirium, convulsions, and unconsciousness. The temperature of the body sometimes rises to a great height, 108°–110° (normal temperature being $98\tfrac{4}{10}°$). The pupils are generally contracted. The pulse at first is strong and rapid, but subsequently becomes weak. Death may occur suddenly in consequence of the action of the high temperature upon the nerve-centers, or, at a later period, as the result of exhaustion. Quite a large percentage of those who recover from the immediate effect of sunstroke subsequently suffer from some temporary or permanent affection of the nervous system.

Those who have been overcome by the heat are peculiarly susceptible to this element for some time afterward.

Heat exhaustion is a mild type of the condition just described and is very common. The symptoms are usually dizziness or faintness, and often some nausea and weakness. These conditions, however, generally respond to simple treatment.

TREATMENT OF HEAT-STROKE.—There are two important indications for treatment—reduction of temperature and the use of stimulants. The use of cold is regarded as the best method of meeting the first indication, and the manner of its application depends upon the facilities at hand. The patient should first be removed to a cooler spot if possible, or at least where shade can be secured, or, if a person succumbs in a place where the ventilation is defective (in addition to the heat), he should be carried to the fresh air. The treatment should be begun at once.

Efforts are often made first to remove the patient to his home, which involves the loss of considerable valuable time and may still further exhaust him. The clothing about the neck and body must be either loosened or removed. Cold is particularly indicated in cases where there is great heat of the body, and delirium and convulsive movements are present. This means of reducing the temperature may be applied in the shape of cracked ice about the head and spine, or the use of cold water, by means of a sponge or by pouring it from a sprinkling-pot, or the patient's head may be held under a pump; however, the best plan is to remove the clothing and sponge the head and body with cold water at frequent intervals. If the heat of the body is very great, it may be necessary to wrap the patient in sheets wet with cold water. The sheets are to be kept wet by frequently pouring water over them while on the body, until consciousness returns or there is an evidence of marked diminution of temperature. After the cold has been discontinued, should serious symptoms, such as unconsciousness or the previous high temperature recur, the cold application should again be employed.

If there is, besides the great heat, evidence of serious depression, stimulants must be used while the cold is being applied.

There are some cases of heat-stroke where stimulants alone are indicated, as when the signs of a high fever are absent and great depression is present. In these cases, in addition to the use of internal stimulation, the application of mustard over the heart and to the calves of the legs is indicated.

The treatment of heat exhaustion consists in rest, bathing the face with an alcoholic solution, or the internal use of a stimulant.

CHAPTER XVI.

ASPHYXIA AND DROWNING.

ASPHYXIA—SUFFOCATION.

Asphyxia is a condition of unconsciousness due to a great diminution of oxygen in the blood, resulting either from an obstruction to the passage of air to the lungs, or to the presence of poisonous gases which render the air unfit for respiration. Among the numerous causes of suffocation are drowning, hanging, and obstruction in the respiratory tract, either by the lodgment of foreign bodies, or by certain maladies affecting this part, as croup. Asphyxia also follows the inhalation of the fumes of charcoal or coke, and the carbonic-acid gas contained in empty wells, caves, beer-vats, and mines; in the latter situation it is known as "choke-damp." Coal-gas from stoves, and illuminating and sewer gases, are frequent causes of suffocation. The appearance of a person suffering from asphyxia is well marked. The face is of a dusky or purplish hue (showing deficient oxygenation of the blood), and swollen. The respirations are extremely labored, and associated with convulsive movements and delirium. If relief is not promptly at hand, these symptoms are rapidly followed by unconsciousness and death.

TREATMENT.—The treatment consists in removing the cause, in order that the lungs may be supplied with the proper amount of pure air, and restoring the different functions to their normal condition by stimulants and artificial respiration. (See DROWNING.)

PRECAUTIONS.—In rescuing a person from an empty

well, care should be taken that the mouth and nose of the person making the descent are protected by holding against them a cloth saturated with water, or vinegar and water. A rope should also be tied around the waist of the rescuer, by which he can be brought rapidly to the open air. Matches should not be ignited while in the well, nor should any artificial light be carried down; for, while carbonic acid gas, which constitutes the bulk of poisonous gases in these receptacles, is not a supporter of combustion, other gases may be present which are ignitible, and a serious explosion would probably follow. Sewer-gases are often inflammable, and illuminating gases always so. A light should never be taken into a cellar or any apartment where gas has escaped, until the room has been thoroughly ventilated by open windows and doors.

It is often necessary to enter empty wells and cess-pools in order to examine or clean them. The carbonic acid which they contain, and which is heavier than air and consequently settles to the bottom, should be stirred up and removed previous to making any descent for examination. This may be done by free ventilation, and by pouring into the well large quantities of water or lime-water, or by lowering and quickly withdrawing an opened umbrella, or throwing down lighted papers and straw. Should the latter means be used, care should be taken not to remain near the opening, as gases may be present which are inflammable. Discharging a gun into the receptacle may also be employed for this purpose.

DROWNING.

The asphyxia or suffocation that follows submersion is due to the fact that air is prevented from reaching the lungs. More or less water is found in the air-passages, but not in such quantities as is generally supposed. In some cases very little if any water reaches these organs, on account of the rapid closure of the epiglottis. Water, however, enters the stomach, and considerable is found mixed with mucus in the throat. Death is usually the result of suffocation, as is made obvious, after the removal of the body from the water, by the bloated and discolored appearance of the face. However, in some cases death may be due to sudden heart-failure before the person sinks. When such is the case the face of the drowned would be pale and flabby, and very little water and mucus would be found in the respiratory tract. There is a better chance of resuscitating one who sinks as the result of syncope than when suffocated, as the demand for oxygen in the former is less than when asphyxiated by submersion.

Persons who are submerged for four or five minutes or more are not usually restored to life; although numerous cases are recorded where resuscitation was effected after an interval of twenty minutes. In such cases it is supposed that syncope occurred, or, on account of the existing excitement, an error was made in calculating the actual time of submersion. The action of the heart usually continues some little time after respiration ceases.

TREATMENT.—The treatment consists first in re-establishing respiration, then stimulating the action of the heart and circulation by the use of stimulants and warmth, friction, etc. When a person has been under water but a few moments, simple means may restore respiration, and should be first tried.

The water, sand, and mucus should first be quickly removed from the mouth and nose, and the attendant should then carry his finger to the back or base of the patient's tongue, which must be pulled forward, thus enabling the water and mucus in the throat and respiratory tract to escape, and also to favor the entrance of air into the lungs; while this is being done the patient should be turned on his side (left, if possible), face downward, to favor the escape of water from the stomach and air passages. He should then again be turned on his back, while the hands of the attendant are placed on the belly or abdomen and pressure directed upward and inward toward the diaphragm. This movement tends to stimulate respiration and should be repeated two or three times at intervals of two or three seconds. The mouth in the mean time should be kept open by a cork or piece of wood, or a knot tied in a handkerchief, etc., in order that the passage of air to the lungs should not be interfered with. Smelling-salts, ammonia, or two or three drops of nitrite of amyl, may be administered by inhalation, or the nose may be tickled with a feather or straw. When breathing commences and consciousness returns, the patient should be carefully divested of all wet clothing (if necessary, the clothing should be cut in order to avoid delay), well rubbed, and wrapped in warm covering, and stimulants given in the manner already described. (SHOCK.)

If the simple measures just enumerated are productive of no good result after a short trial, artificial respiration should be at once resorted to.

Three methods of artificial respiration are accepted as being particularly efficient, viz., *Sylvester's*, *Howard's*, and *Hall's*. The first (Sylvester's) is generally regarded as being the best. The methods of Sylvester and Howard may be performed by one person, which is an important consideration.

Before artificial respiration is begun, the patient should be stripped to the waist, and the clothing around the latter part should be loosened so that the necessary manipulations of the chest may not be interfered with.

SYLVESTER'S METHOD.—The water and mucus are supposed to have been removed from the interior of the body by the means above described. The patient is to be placed on his back, with a roll made of a coat or a shawl under the shoulders; the tongue should then be drawn forward and retained by a handkerchief which is placed across the extended organ and carried under the chin, then crossed and tied at the back of the neck. An elastic band or small rubber tube or a suspender may be substituted for the same purpose. If no other means can be made available, a hat- or scarf-pin may be thrust vertically through the end of the tongue without injury to this organ. The attendant should kneel at the head and grasp the elbows of the patient and draw them upward until the hands are

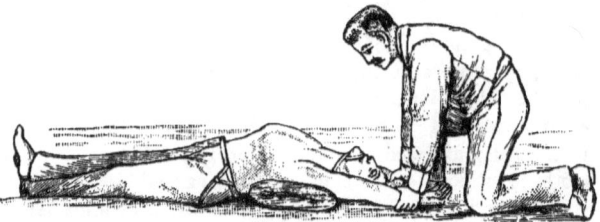

FIG. 80.—Sylvester's method. First movement (inspiration).

carried above the head and kept in this position until one, two, three, can be slowly counted (Fig. 80). This movement elevates the ribs, expands the chest, and creates a vacuum in the lungs into which the air rushes, or, in other words, the movement produces *inspiration*. The elbows are then slowly carried downward, placed by the side, and

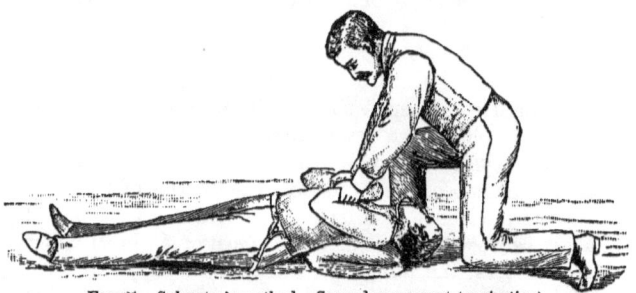

FIG. 81.—Sylvester's method. Second movement (expiration).

pressed inward against the chest, thereby diminishing the size of the latter and producing *expiration* (Fig. 81). These movements should be repeated about fifteen times during

DROWNING.   183

each minute for at least two hours, provided no signs of animation present themselves.

HOWARD'S METHOD is divided into two parts. *Part first* consists in removing the water, etc., from the respiratory tract and stomach, and is as follows:

The patient should be placed face downward, with a pillow or roll of clothing under the pit of the stomach, the head resting on the forearm, which keeps the mouth from the ground and renders traction on the tongue unnecessary. The attendant, standing over the drowned person, should then place his left hand on the lower and back part of the left side of the chest, while the right hand is laid on the

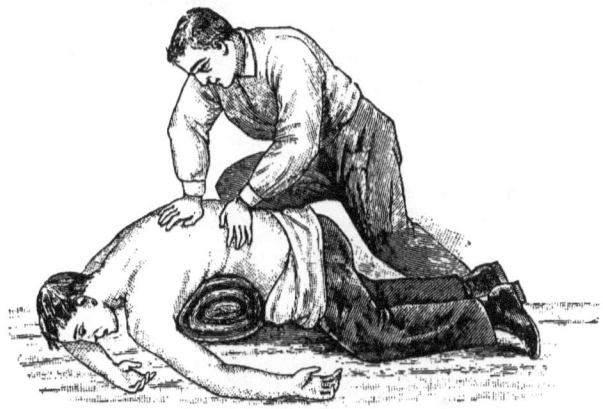

FIG. 82.—Howard's method. Part first.

spinal column about on a line or a little above the left hand; firm pressure is then made by the operator throwing the weight of his body forward on his hands; this is to be continued while one, two, three, are counted (slowly), and ended with a push which helps to raise the operator to an upright position and forcibly expel the fluid. These movements should be repeated two or three times if fluid continues to flow from the mouth (Fig. 82).

Part Second : Artificial Respiration.—The patient should now be turned on his back, face directly upward, with the roll placed under the shoulders; it should be made thick enough to allow the head and neck to be fully thrown back. The hands of the patient should be carried above his head and tied with a handkerchief, suspenders, etc. The attendant should then kneel astride the patient's hips and place the palm of his open hands upon the lower part of the patient's chest with the thumbs at the lower border of the breast-bone, the fingers being applied to the spaces between the ribs, the little finger being laid along the lower border of the ribs, in front. The hands are then pressed slowly and firmly upward and inward toward the diaphragm or midriff, with the body of the operator thrown forward until his face is nearly in contact with that of the patient. A sharp push is then made upon the chest, which helps to bring the operator to an upright position. A rest of two or three seconds should then follow, and the movement repeated. These procedures should be continued for about two hours before being abandoned, unless the patient is sooner restored (Fig. 83).

Fig. 83.—Howard's method. Part second.

HALL'S METHOD.—The drowned person should be placed face downward, the head resting on his forearm with a roll or pillow placed under the chest; he should then be turned on his side, an assistant supporting the head and keeping the mouth open; after an interval of two or three seconds, the patient should again be placed face downward and allowed to remain in this position the same length of time. This operation should be repeated fifteen or sixteen times each minute, and continued (unless the patient recovers) for at least two hours (Figs. 84, 85).

FIG. 84.—Hall's method. First position.

If, after using one of the above methods, evidence of recovery appears, such as an occasional gasp or muscular movement, the efforts to produce artificial respiration must not be discontinued, but kept up until respiration is fully established. All wet clothing should then be removed,

FIG. 85.—Hall's method. Second position.

the patient rubbed dry, and if possible placed in bed where warmth and stimulants can be properly administered. A small amount of nourishment, in the form of warm milk or beef-tea, should be given, and the patient kept quiet for two or three days.

CHAPTER XVII.

POISONS AND POISONING.

CERTAIN agents which are invaluable in the treatment of the sick and injured become deadly poisons when improperly used. This may be the result either of accidents, errors in administration, or attempts at self-destruction.

Poisons may be divided into *narcotics* and *irritants*. Some of the irritants are exceedingly destructive to the tissues with which they come in contact, and are known as *corrosive* poisons.

Narcotics are employed in the practice of medicine to produce a soothing effect or sleep. When these drugs are taken in poisonous doses, they are followed by symptoms which in a general way are common to all of this class, and by which they can be recognized in cases of poisoning, viz., delirium, stupor, insensibility, and stertorous breathing. Narcotics cause very little, if any, local irritation.

Irritants are used medicinally both for a constitutional and local effect. Their poisonous action is particularly noticeable by the local irritation they produce. Great destruction of the tissues with which they come in contact follows the use of the corrosives. This is markedly apparent about the lips, mouth, and throat. Intense abdominal pain, associated with vomiting and purging, and also shock, are more or less constant.

TREATMENT OF POISONING.—The treatment of poisoning is both local and constitutional.

The local treatment consists in preventing the further

absorption and irritating effect of the poison by removing it from the stomach by the use of emetics and the stomach-pump. Remedies are also used which in a measure render the operation of the poison harmless (antidotes).

The constitutional treatment is directed toward antagonizing the action of the poison, and stimulating the patient.

Agents which directly counteract the local and constitutional effect of a poison are known as antidotes: for example, when an acid is taken in a poisonous dose, an alkali, such as soda, is given as the antidote; or, when opium is the poison introduced, belladonna is administered as one of the physiological antidotes. These examples demonstrate local and constitutional antidotes.

LOCAL TREATMENT.—EMETICS.—Under this heading may be included the introduction of the forefinger or a feather into the throat, and by passing either backward as far as possible can be employed to induce vomiting (emesis). This means may also be used to stimulate the action of emetics which have been given internally.

Mustard.—One tablespoonful added to half a pint of tepid water. Mustard is a stimulant emetic, and very valuable in narcotic poisoning.

Chloride of Sodium (common salt).—Two tablespoonfuls added to half a pint of tepid water. Not very certain as an emetic, but generally on hand.

Alum.—Tablespoonful in half a pint of tepid water. Rather a feeble emetic.

Sulphate of Zinc (white vitriol).—Twenty to thirty grains in half a glass of water. Very efficient and safe. One of the best emetics.

Ipecac.—Thirty grains of the powder or two tablespoonfuls of the wine or sirup of ipecac; the latter preparation is apt to be found in the house.

Carbonate of Ammonia.—Thirty grains in half a glass of water.

Sulphate of Copper (blue vitriol, blue-stone).—Five to ten grains in half a glass of water.

The dose recommended of each of the above emetics is for adults, the amount should be proportionately small for children.

The contents of the stomach may be removed by siphonage. This method is simple and effective, and is performed as follows: One end of a small rubber tube about six feet long, after being oiled, is passed into the stomach. This procedure requires considerable care, as the tube may be introduced into the air-passage by mistake, instead of into the œsophagus; for this reason the tongue should be depressed and the end of the tube carried backward as far as possible and then downward. After this has been accomplished, the end projecting from the mouth is raised and a funnel attached, through which tepid water (about two or three pints) should be poured into the stomach. The upper end of the tube (still raised) should be pinched between the thumb and finger, and in this condition lowered below the level of the stomach. When the thumb and finger are removed, the water held in the tube acts as a siphon, and the contents of the stomach are withdrawn, provided the poisoning does not occur soon after a meal. This operation may be repeated a number of times, until the poison has been removed. It is always best, particularly in cases of suicide, to place a cork, or, still better, a spool of thread, as a gag between the teeth, to prevent the tube from being bitten in two.

The use of a stomach-tube is *not* indicated when the poisoning is due to *corrosives*, as the mucous membrane is greatly swollen, and the introduction of the tube is productive of great pain and danger of laceration of the soft tissues. The ordinary fountain syringe, which is found in almost every house, is an excellent instrument for washing out the stomach. The rubber or metal end being cut off, the tube can be introduced. The attached fountain makes the use of a funnel unnecessary.

CONSTITUTIONAL TREATMENT.—The constitutional antidotes will be referred to with the special poisons. Stimulants are indicated to overcome the depressing action of the different poisons, particularly the narcotics, and may be administered by the mouth (if the patient can swallow), provided the poisoning is not due to *corrosives*, or by the rectum (lower bowel), or by inhalation. The stimulants administered are those ordinarily used, and in addition strong coffee; the latter is of great value in narcotic poisoning, and it may be given in the same manner as other stimulants, either by the mouth or rectum. When stimulants are given by the rectum, the amount should be considerably larger than when given by the mouth, and they should be diluted with sufficient warm water or milk to make a cupful. Ammonia, and nitrite of amyl can be given by inhalation: they should always be administered by dropping a sufficent amount on a handkerchief or the palm of the hand. The bottle should never be held to the nose, as the contents might be spilled into the mouth or nasal cavity. Stimulants should be given by inhalation with great caution, where corrosive poisons have been taken, as they are liable to irritate still more the inflamed membrane. Nitrite of amyl, which is given by inhalation, is a powerful heart stimulant, but should be used with care and in small amounts—three to five drops. This remedy is now obtained in small glass capsules, which contain about five drops, and are broken when the nitrite of amyl is needed.

The following list of narcotics and irritants comprises those which are commonly the cause of poisoning.

NARCOTICS.

ACONITE.—(Monk's-hood, wolf's-bane.) The tincture has the appearance of sherry wine, for which it has been mistaken. Aconite is commonly found in fever-mixtures, neuralgia "cures," liniments and ointments, and is exceedingly poisonous.

Symptoms.—Great prostration; tingling of the lips, mouth and throat, and extremities—the latter become numb; slow and weak pulse; shallow respiration; the mind is generally clear; convulsions may sometimes occur.

Treatment.—Emetics; free use of stimulants; warmth to extremities; twenty drops of tincture of belladonna, repeated (once) in half an hour; mustard over heart and calves of legs; artificial respiration.

ALCOHOL.—Acute poisoning is very uncommon. A large amount is sometimes taken for a wager, or by one unaccustomed to its use.

Symptoms.—(See INTOXICATION.)

Treatment. — Keep the patient aroused by pinching, slapping with wet towel, or hot and cold douches to head and spine; coffee in large amount by stomach or rectum; inhalation of ammonia, and artificial respiration.

BELLADONNA. — (Deadly night-shade.) Contained in cough-mixtures, etc., liniments, ointments, and plasters. *Atropine*, the active principle of belladonna, is commonly found in eye medicaments. Poisoning sometimes follows the use of an external application containing either belladonna or atropine, it being absorbed through the unbroken skin.

Symptoms. —Headache; intense dryness of mucous membrane of mouth, throat, and nasal cavity; inability to swallow; eyes brilliant (belladonna is often used for this purpose); *pupils dilated;* face usually flushed; delirium and stupor, or convulsions. Pulse accelerated.

Treatment. — (Belladonna and atropine.) Emetics; thirty drops of laudanum, or one to two tablespoonfuls of paregoric, or one quarter of a grain of morphine, repeated (once) in half an hour; stimulants; warmth to extremities; artificial respiration.

CAMPHOR.—Usually taken in the form of a tincture (spirits), or the gum.

Symptoms.—Characteristic odor; surface of body cold and clammy; disturbances of vision; noise in ears; delirium and convulsions.

Treatment.—Emetics; stimulants; warmth to extremities; hot and cold douches.

HYDRATE OF CHLORAL. — (Chloral.) Odor similar to bananas or pears. Chloral is often an ingredient in sleeping potions, cough mixture, and liniments.

Symptoms.—Face congested; weak pulse; labored respirations, stertorous breathing; stupor and unconsciousness, sometimes convulsions.

Treatment.—Free use of stimulants; warmth to extremities; mustard over heart and calves of legs; artificial respiration. An emetic may be given, if the patient is seen immediately after a poisonous dose has been taken.

CHLOROFORM.—When poisoning is caused by inhalation, the patient's tongue should be drawn forward and artificial respiration performed; hot and cold douches; also friction; plenty of fresh air; the patient's head should be low, or it may be allowed to drop down, the body being elevated by carrying the legs over the shoulder of the attendant.

When poisoning follows the internal use of chloroform, large doses of bicarbonate of soda in water should be given internally, in addition to other means of resuscitation.

DIGITALIS.—(Purple foxglove.) Generally given for heart maladies.

Symptoms.—Purging and vomiting; great pain in abdomen; face pale; pupils dilated; skin cold and clammy; pulse feeble and slow; great depression; gasping respiration; delirium, and convulsions.

Treatment.—Emetics (only in case one large dose of digitalis has been swallowed); about twenty grains of tannin in solution, or white-oak bark, or strong tea, repeated; mustard, especially over the heart, and the calves of the legs, and stimulants internally. The patient should be kept in the recumbent position for some time after the serious symptoms have ceased.

FUNGI.—Poisonous mushrooms.

Symptoms. — Colic; vomiting and purging; dilated pupils; extreme muscular weakness; mental excitement. Symptoms usually occur within an hour.

Treatment.—Emetics; castor-oil; stimulants; heat. Atropia is a physiological antidote; $\frac{1}{100}$ of a grain may be given and repeated once. "It is always dangerous to warm up a dish containing mushrooms." (Murrel.)

The treatment of mushroom-poisoning is applicable to poisoning by eating mussels.

HYDROCYANIC ACID.—(Prussic acid.) The dilute acid is the form in which this poison is generally used, and is a colorless fluid. Prussic acid is one of the deadliest poisons known. The poisonous symptoms begin almost immediately.

Symptoms.—Great and sudden prostration; surface cold and clammy; pupils dilated; eyes fixed; blueness of face, locked jaws; gasping respiration and covulsions.

Treatment.—Dilatory treatment is of no avail in poisoning by this agent; whatever is to be done must be done at once; no time should be lost in trying to give emetics, or use the stomach-pump; stimulants should be given freely; flapping with wet towel; hot and cold douches; artificial respiration. In poisoning by this drug, stimulants should be given by hypodermic injection, even by a non-professional person; a teaspoonful of brandy or whisky, or ammonia-water (not spirits), introduced into the calf of the leg, and repeated in five minutes; inhalations of nitrite of amyl.

The symptoms of poisoning which may occur after cyanide of potassium, peach-pits, and bitter almonds have been taken internally are dependent upon the presence of prussic acid, and should be treated accordingly.

OPIUM.—The preparations generally used are morphine (the active principle of opium), tincture of opium (laudanum), *camphorated* tincture of opium (paregoric), Mc-

Munn's elixir, codeine, and narceine. Opium is an active ingredient in many preparations, particularly sedative and cough mixtures, soothing syrups for children, and numerous patent medicines.

Symptoms.—Usually begin with more or less mental excitement, followed by great depression, headache, sensation of weight in the limbs, an irresistible desire to sleep, which soon deepens into a stupor; *the pupils are contracted* and the respirations are *greatly diminished in number*, sometimes to four or five a minute, and the skin is cold and clammy and the face pale. Convulsions sometimes occur.

Treatment.—Emetics; cold douche to head, face, and spine; flapping with wet towel; coffee by rectum or mouth. Wines and liquors *must not be given.* Twenty drops of the tincture of belladonna repeated (once) in fifteen minutes. Artificial respiration for two or three hours. The patient should be carefully watched for some time after the dangerous symptoms have subsided.

IRRITANTS.

ARSENIC.—This agent is used medicinally, generally in the form of Fowler's solution and arsenious acid; it is also contained in rat-pastes, fly and vermin killers, and is employed to a large extent in the arts, for coloring flowers, wall-paper, stuffing birds, etc. Realgar is a red, and Orpiment, or King's yellow, is a yellow preparation of arsenic. Arsenite of copper (Paris green, or Scheele's green) is commonly taken by suicides.

Symptoms.—Intense burning pain in pit of stomach; faintness and depression; vomiting and purging; vomited matter streaked with blood. Abdomen tender and painful; skin cold and clammy. If Paris green has been taken, the green color is generally apparent about the mouth, hands, or clothing.

Treatment.—Emetics; large draughts of hot, greasy water, or salt and water; dialyzed or tincture of iron; a large amount of magnesia may be given; or lime which may be scraped from the walls or ceiling or from a fence; castor-oil; sweet-oil, or equal parts of sweet-oil and lime-water, or lime-water alone; raw eggs; milk; stimulants (well diluted if given by the mouth); twenty to thirty drops of laudanum, or one quarter of a grain of morphine, to allay the pain.

CANTHARIDES.—(Spanish fly.)

Symptoms.—Burning sensation in mouth and throat; violent pain and soreness in abdomen; vomiting and purging, vomited matter containing streaks of blood; great irritation of urinary organs and passage of bloody urine; delirium and convulsions.

Treatment.—Emetics, raw eggs, sweet-oil, milk, laudanum (twenty drops), or morphine (one quarter of a grain), may be given. Stimulants (by rectum), if necessary; heat; poultices over abdomen.

COPPER, SULPHATE OF (blue vitriol, blue-stone), or subacetate of copper (verdigris), is sometimes the cause of poisoning, although copper-poisoning is more frequently due to substances which have been cooked in copper vessels, which is an unsafe method of preparing food.

Symptoms.—Vomiting and purging; griping pain in abdomen; metallic taste in mouth; great thirst and weakness; labored respirations and rapid pulse; dimness of vision; and sometimes convulsions.

Treatment.—Copious draughts of warm water; emetic; raw eggs; milk; stimulants; morphine (one quarter of a grain); laudanum (twenty to thirty drops).

CROTON-OIL.—Has been mistaken for castor-oil.

Symptoms.—Vomiting; purging; violent pain in abdomen; great depression; skin cold and clammy.

Treatment.—Emetics; raw eggs; stimulants; spirits of camphor, ten drops in water every five minutes until five or six doses have been taken; morphine (one quarter of a grain); laudanum (twenty to thirty drops); paregoric (one to two tablespoonfuls).

Iodine.—As generally used is in the form of the tincture—a dark reddish fluid.

Symptoms.—Great local irritation and pain; the mucous membrane of the mouth is discolored yellow; vomiting and purging, the vomited matter having a *bluish* color, as the result of the admixture of the iodine and starchy matter of the food (iodide of starch). *Blue* color also noticed when vomited matters come in contact with starched articles of clothing.

Treatment.—Free use of starch, then emetics, raw eggs, chalk, magnesia, and stimulants.

Strychnine.—(The active principle of nux-vomica).

Symptoms.—An irritable condition of the muscular system, and twitchings and discomfort, soon followed, usually within less than an hour, by severe convulsions, which occur at intervals of two minutes to half an hour. The contractions of the powerful muscles of the back cause the body to assume the form of a bow, the patient resting on the head and heels (opisthotonos); great pain is present as the result of the intense muscular contraction. The mind is usually clear. The respiratory muscles are prominently involved in this affection, and death by asphyxia, as the result of rigidity of these muscles, may follow, or death may occur later, as the result of exhaustion.

Strychnine-poisoning can only be mistaken for tetanus (locked-jaw). The distinction is to be made as follows: In strychnine-poisoning the attack is sudden; there is generally evidence of poison having been taken, and no injury has been received. In tetanus, the development of the disease is more gradual; no poison has been taken, but some form of injury is present. In strychnine-poisoning the muscles of the jaw are only slightly affected, if at all, but the upper extremities are always involved; while in tetanus, the rigidity of the muscles of the jaws (locked-jaw) is the first and prominent symptom, and continues to the end, the upper extremities are only affected in the severe cases, and then only slightly. In strychnine-poisoning, the duration of the attack covers but a few hours, while in tetanus a number of days may be involved.

Treatment.—The stomach-tube can only be used if the patient is seen immediately after the poison is taken, as the convulsive movements prevent the introduction of the tube; a mixture of tannin or animal charcoal, and followed by an emetic; then by large doses of bromide of sodium or potassium (sixty grains in solution), repeated every hour until three or four doses have been taken, together with nitrite of amyl; artificial respiration.

PHOSPHORUS.—Poisoning frequently occurs as the result of sucking or swallowing the heads of matches. This substance is also contained in "rat pastes."

Symptoms.—Vomiting and purging; vomited matter has a garlicky odor, and is luminous in the dark. Intense pain in abdomen.

Treatment.—Emetics; free use of magnesia; French turpentine, given in an alcoholic or alkaline solution, half teaspoonful every half-hour, until a number of doses have been taken (five or six): three grains of sulphate of copper every five minutes until free vomiting occurs; white of egg.

TARTAR EMETIC.—Antimony, or Stibium.

Symptoms.—Vomiting and purging; vomited matter contains streaks of blood; skin cold and clammy; collapse.

Treatment.—Warm water; mixtures of tannin; white-oak bark; also strong tea; stimulants, and warmth.

ZINC.—The sulphate or chloride is generally used. The chloride is far more dangerous and corrosive.

Symptoms.—Pain and vomiting, with great depression, and, if the chloride has been taken, the mucous membrane of the mouth is inflamed.

Treatment.—Large quantities of bicarbonate of soda (baking-powder) or carbonate of potash in solution; milk; eggs; mixtures of tannin or strong tea, or oak-bark; morphine (one quarter of a grain); laudanum (twenty to thirty drops).

CORROSIVES.

ACIDS.—Sulphuric (oil of vitriol); muriatic (spirit of salt); nitric (aquafortis); acetic; carbolic, and oxalic acids.

Symptoms.—Mucous membrane of lips, mouth, and throat greatly inflamed and swollen, with more or less destruction of the tissue, and symptoms of shock. When taken in a concentrated form, muriatic and sulphuric acids discolor the lips and mouth *black*, while nitric acid stains these tissues *yellow;* when oxalic and carbolic acids have been taken, the membrane has a *whitish* appearance.

Treatment.—For sulphuric, muriatic, nitric, and acetic acids; immediate use of alkalies, as mixtures of soda, potash, magnesia, chalk, white crayons, lime, soap-suds, or tooth-powder (chalk); raw eggs, milk, oil. Rectal stimulation for shock.

OXALIC ACID.—Is sometimes mistaken for Epsom salts. Is used for cleaning kitchen boilers, and sometimes introduced accidentally into food.

Symptoms.—Same as other corrosive poisons.

Treatment.—Magnesia or lime in some form should be administered—the latter may be scraped from the wall or ceiling—white crayons; tooth-powders, etc. Soda, potash, and ammonia should *not* be used. Castor-oil or sweet-oil should be administered by the mouth, and rectal stimulation resorted to. Poisoning by salts of lemon and sorrel, which are derivatives of oxalic acid, should be treated as above.

CARBOLIC ACID, CREASOTE (Phenol).

Symptoms.—Same as other *corrosives*, and also distinguished by their characteristic odor.

Treatment.—Alkalies, particularly sulphate of magnesia (Epsom salts) or sulphate of soda (Glauber's salts) given in large doses; raw eggs; castor-oil, and sweet-oil. Emetic if necessary.

CORROSIVE SUBLIMATE.—(Bichloride of mercury.) Used largely as an antiseptic, and to preserve specimens, etc.

Symptoms.—Similar to those of arsenical poisoning. Vomiting and purging ; intense pain in abdomen ; membrane of mouth and tongue white and shriveled ; metallic taste in mouth.

Treatment.—Emetics, if vomiting is not present. White of egg in large amount ; milk or mucilage in abundance ; flour and water or arrowroot or barley-water ; chlorate of potash. Stimulants by rectum.

ALKALIES.—Caustic potash and soda ; pearlash ; strong solution of ammonia.

Symptoms.—Similar to poisoning by corrosive acids. Intense swelling and inflammation of lips and mouth.

Treatment.—The treatment consists in the administration of *acids*—vinegar, lemon or orange juice, hard cider; also acetic, nitric, muriatic, and sulphuric acids *largely* diluted ; raw eggs ; olive-oil ; milk ; arrowroot, and barley-water. Stimulants by the rectum. Morphine (one quarter of a grain) or laudanum (twenty to thirty drops) may be given.

POISON IVY OR OAK. *Symptoms.*—A vesicular eruption of a brownish-red color appears; this may be associated with the formation of large blebs or blisters; the skin becomes very red, swollen, hot, and irritable, with intense burning and itching.

Treatment.—As the poison is due to a volatile acid, an alkaline application is indicated ; for this purpose a solution of baking soda or saleratus may be used, or, still better, strong soap-suds. Later, an application of a solution of the acetate, or "sugar" of lead, thirty to sixty grains to a pint of water, is beneficial. Dry starch dusted over the affected part is also used at this stage.

CHAPTER XVIII.

THE CONVULSIONS OF CHILDREN.

CONVULSIONS affecting children are of common occurrence, but do not carry with them the gravity which is associated with a like affection in adults. The convulsions are often the result of trivial causes, as indigestion, teething, worms, swallowing foreign bodies. They may sometimes, however, usher in serious affections, as scarlatina (scarlet fever). At the time of the convulsion the thumbs are carried across the palm of the hand; the eyes are set and turned upward; then suddenly the patient becomes rigid, with some contraction of the facial muscles; the face and lips become dusky, as the result of considerable interference with respiration. This rigidity lasts but a few moments, and is generally followed by a stupor, which often terminates in a natural sleep and recovery.

Treatment.—The head should be enveloped in a cloth wet with cold water, and the child at once placed in a hot bath (about 100°). A tablespoonful of mustard may be added to the water, the efficiency of the bath being thereby increased. The child should remain in the bath about two or three minutes, and should be again immersed if the convulsion is repeated.

TETANUS—LOCKED-JAW.

Tetanus is a disease characterized by spasmodic rigidity of the voluntary muscles, and is sometimes the result of a wound of the foot or hand, made by a blunt or jagged implement, as a rusty nail. Tetanus may also follow burns. Cases of the disease are not frequent. An interval of from five to fifteen days generally follows the reception of the injury before the symptoms become apparent.

The first well-marked symptoms of the disease are pain and stiffness about the neck and jaws, which is commonly mistaken for rheumatism. This increases in severity, and the muscles of mastication become involved and the patient is unable to open his mouth (trismus). The name by which the disease is commonly known (locked-jaw) is due to this fact. The rigidity about the neck soon extends to other portions of the body. The strong muscles of the back being so forcibly contracted, the body is bent in the shape of a bow and rests on the head and heels (opisthotonos). The upper extremities are but slightly affected in mild cases. The contraction of the facial muscles elevates the corners of the mouth, and produce a characteristic appearance known as risus sardonicus, or canine laugh. This affection is usually followed by a fatal result.

Tetanus is to be distinguished from strychnine-poisoning. (See STRYCHNINE-POISONING.)

Treatment.—Medicine is of very little use in tetanus; however, large doses of bromide of potassium (thirty to forty grains) may be given with safety three or four times daily). Operative measures are often undertaken by the surgeon. The patient's strength should be kept up as well as possible by nourishment, such as milk, beef-tea, and kumyss, which may be carefully poured into the mouth between the teeth and cheeks; in this way it reaches the stomach. Excitement of all kinds should be avoided.

FOREIGN BODIES IN THE EYE, EAR, NOSE, LARYNX, AND PHARYNX.

EYE.—Foreign bodies in the eye usually consist of cinders, sand, dust, and minute scales of iron. The latter have become a frequent source of annoyance since the elevated railroads have been in operation.

Treatment.—The foreign bodies just described are usually lodged about the center and under surface of the upper lid, and are often washed out of the eye by the increased flow of tears which follows their introduction. If the foreign body still remains lodged, the upper lid should be gently pulled away from the eye with the thumb and finger, and carried downward and pressed against the lower lid, to which the particle may be transferred. Should this not be successful, the upper lid may be turned upward and folded on itself over a pencil, when the foreign body can generally be seen and removed by lightly touching it with a small fold of a clean handkerchief or other similar substance. The operation of eversion, or turning the lid upward, usually requires considerable practice before it can be done dexterously. The lower edge of the upper lid should be carried away from the eye and somewhat upward by seizing the lid with the thumb and forefinger of the left hand—the operator standing in front of the patient—while with the tip of the index-finger of the other hand, or the shaft of a lead-pencil, the center of the lid is pressed downward, and the inner surface is exposed. The particles of iron already referred to, and also some other foreign bodies, have sharp cutting edges, and often penetrate the eyeball, usually over the pupil, and can not be dislodged by the ordinary process. All unskillful manipulation, and also the common practice of rubbing the eye, press the substance still deeper into the structures. This condition should, if possible, be attended to by a surgeon. The pain and reddened condition of the eye may be relieved by applications of warm water, tea, or rose-water, or two or three drops of a four-per-cent solution of cocaine dropped in the inner corner of the eye.

EAR.—Greater danger, probably, results from the unskillful, unnecessary, and rough manipulation, in cases where foreign bodies have been introduced into the ear, than from the presence of the article itself. A foreign body, even of large size, which does not cause immediate pain and distress, may remain in the ear for some time without causing any serious result. It can, at least, remain until removed in a proper manner. Only the simplest and gentlest measures should be employed by a non-medical person. Among the foreign bodies which are found in the ear are insects, bugs, peas, beans, pebbles, cherry-pits, and shoe-buttons. Children commonly introduce the latter articles into their own ears or the ears of playmates.

Treatment.—Bugs and insects are readily removed by carefully syringing the affected ear with warm water; or, if a syringe can not be procured, then water can be poured into the ear. This procedure either drowns the insect or drives it out, and is so uniformly successful and simple that it is regarded as the best treatment for this purpose. Another method, which is commonly employed, consists in applying a piece of cotton saturated with salt water or vinegar to the ear, and when the cotton is removed, after an interval of a few minutes, the insect is usually found attached to it. Gentle syringing should also be employed to remove other foreign bodies from the ear; *exception should, however, be made when the body is liable to be increased in size by the use of water, as peas and beans.* Tooth-picks, button-hooks, and other similar instruments, should *never* be used; for when they are employed, the foreign body is almost always pushed further into the canal, and its subsequent extraction by the surgeon made more difficult, and the drum of the ear is very liable to be perforated, or other unpleasant accidents ensue.

NOSE.—Foreign bodies are often inserted into the nose by children and insane persons, or as the result of accident. The article introduced may be small and not particularly irritating, and may remain in the nasal cavity for a long time before its presence is known.

Treatment. — Tickling the opposite nostril with a feather until sneezing is induced, is very often all that is necessary, particularly if the foreign body has recently been inserted; if not successful, the nasal douche of warm water should be employed through the unaffected nostril, unless the foreign body is likely to be increased in size by contact with water. The reservoir which holds the water should be raised above the head, in order that a continuous stream may be secured. If the mouth is kept open, the water will better escape from the affected nostril. If the simple means just described are not followed by the desired result, the case should be seen by a surgeon.

LARYNX.—Foreign bodies may enter the larynx during a fit of laughing, or when a sudden act of inspiration is made; this occasionally occurs while eating. A piece of meat is often a cause of strangulation. The obstruction to the entrance of air into the lungs is immediately followed by a violent paroxysm of coughing, which usually expels the foreign body; if not dislodged, evidences of suffocation immediately follow. The face becomes congested and the eyes protrude; the inspirations are gasping in character, and the patient makes frantic efforts for relief. The substance causing the obstruction is usually in such a position that some air reaches the lungs.

Treatment.—The *nearest* surgeon should *at once* be sent for, and always informed as to the character of the case, in order that he may bring with him the instruments necessary for an operation, which usually consists of an external opening into the windpipe. In the mean time the subject should be slapped violently on the upper portion of the back between the shoulder-blades, particularly at the moment of coughing. If this is not successful, invert the body—that is, hold the patient by his feet with the head down—and again slap the back. This should at once be discontinued if it increases the distress. If the accident occurs during a meal, the finger should be passed into the throat, with a possibility of dislodging the piece of meat or whatever may cause the mischief.

If the patient becomes unconscious from the consequent asphyxia, artificial respiration should be employed until the surgeon arrives. Fresh air, or the use of oxygen contained in cylinders, is of value in the treatment.

PHARYNX.—The local distress caused by foreign bodies

in the pharynx—which usually consist of articles of food, coins, fish-bones, pins, needles, false teeth—depends upon their size, which, if large, produce more or less pressure over or upon the air-passage, together with symptoms similar to those of foreign bodies in the larynx; or, if sharp-pointed, the article may cause great pain and irritation.

Treatment.—Foreign bodies in the pharynx are often removed by acts of vomiting or coughing; if not so expelled, an examination of the throat should be made, and, if possible, the substance removed ; if this can not be done, and the foreign body is not too large or sharp, it may be pushed downward into the stomach. A fish-bone, pin, or needle is almost always found sticking into the mucous membrane, and care should be observed, while trying to remove it, that it is not forced still further into the tissues.

Children often swallow coins, buttons, etc., and sometimes substances of very large bulk. When this occurs, emetics and cathartics should *not* be given. Emetics are objectionable, because it is safer to have the foreign body pass downward through the alimentary tract. Instances have occurred where coins, etc., have been brought up by the use of emetics and have entered the air-passages before they could be expelled from the mouth. Cathartics should not be used, for the reason that they irritate the intestinal tract without doing any good. The child should receive the usual food, with the addition of wheat or rye flour in the form of a gruel; this material adheres to the foreign body, and prevents it from irritating the canal on its way down. After a foreign body, such as is usually swallowed, reaches the stomach, it is rarely followed by a serious result.

The passages from the bowel should always be carefully examined for the substance, which usually makes its appearance within three days. The medical attendant must be at once informed if the child becomes restless or feverish, or is affected with pain.

The following very valuable remarks, in regard to foot-soreness, are made by Prof. Parkes (Parkes' "Manual of Hygiene"):

"Foot-soreness is generally a great trouble, and frequently arises from faulty boots, undue pressure, chafing, riding of the toes from narrow soles, etc. Rubbing the feet with tallow or oil, or fat of any kind, before marching, is a common remedy. A good plan is to dip the feet in very hot water before starting, for a minute or two; wipe them quite dry, then rub them with soap (soft soap is the best) till there is a lather; then put on the stockings. At the end of the day, if the feet are sore, they should be wiped with a wet cloth, and rubbed with tallow and spirits mixed in the palm of the hand (Galton). Pedestrians frequently use hot salt and water at night, and add a little alum. Sometimes the soreness is owing simply to bad stockings; this is easily remedied. Stockings should be frequently washed, then greased. Some of the German troops use no stockings, but rags folded smooth over the feet. This is a very good plan.

"Very often the soreness is owing to neglected corns, bunions, or in-growing nails. If blisters form on the feet, the man should be directed not to open them during the march, but at the end of that time to draw a needle and thread through the blister; the fluid then oozes out.

"All foot-sore men should be ordered to report themselves at once."

CHAFING.

Chafing is particularly common among soldiers, and is generally due either to the friction caused by badly fitting clothes or where surfaces of the skin rub together, as the inside of the thigh, or at the flexures or bendings of the different joints of the body.

Treatment.—The cause of irritation should be removed if possible, and the affected parts washed with cold water, dried thoroughly, and then dusted with fuller's earth, bismuth, lycopodium, oxide of zinc, starch, oatmeal, or flour. Fuller's earth is superior to the others mentioned for this purpose; flour and starch are apt to become sour, and increase the local irritation.

Unless the affection is promptly attended to, inflammation of the skin may follow, which requires rest together with active and prolonged treatment.

CHAPTER XIX.

HYGIENE.

THE subject of hygiene, which relates to the care and preservation of health and prevention of disease, may very properly be introduced by indicating in a general way the manner in which the body should be cared for. In order to intelligently understand this matter, a general knowledge of anatomy and physiology is necessary.

It will be seen that the skin is an exceedingly active organ, constantly throwing off the perspiration, which contains considerable waste matter; in addition, the sebaceous follicles supply an oily substance for lubrication; these, with the worn-out, scaly, or superficial layer of the skin, produce a large amount of effete or decomposed material, which, if not promptly removed, embarrasses the function of this organ and favors the formation of unpleasant odors. It will thus be appreciated that cleanliness is essential to health.

Baths.—Considerable has been written upon the subject of baths, and numerous and complicated methods have been advocated. These in a great measure are luxuries, and are not necessary to the well-being of the body. A cold bath upon rising, the water to be applied by the hand and followed by moderate friction, is, as a rule, sufficient to keep the skin properly cleansed and stimulated. This applies to healthy adults, and is not recommended for children and old people. In the latter subjects a warm bath at bedtime, two or three times weekly, will suffice. Plunging into a tub of cold water, particularly in the winter time, is of questionable value, and can only be done with safety by the strongest. If the desired effect is obtained, the use of cold water should be promptly followed by a glow or reaction.

Warm baths during the day should be followed by the application of cold water, which renders a person less susceptible to exposure. The temperature of a bath can be changed without any ill effects. Baths (except those upon rising in the morning) should not be taken just before or after a meal. Cold baths should not be taken when hungry or fatigued.

Warm baths are soothing, and allay nervous irritability.

While Russian and Turkish baths are often valuable for therapeutical purposes and as extreme cleansing agents, they are rather sources of pleasure, and are not essential to the welfare of the skin. Sea-bathing is usually overdone, and often followed by unsatisfactory results. Persons suffering from head and chest affections should not go into the sea. Adults should not remain in the surf longer than fifteen or twenty minutes, and children and old and delicate subjects even less. Noon is perhaps the best time of day for sea-bathing, and the bath should not be taken close to meal-time. The bather should not wait to cool off, but plunge in at once and immerse the whole body; when warm, a person can better stand the shock of the cold water. After the bath the body should be quickly dried, and a brisk walk taken.

The addition of alcohol, cologne, ammonia, etc., make the bath particularly pleasant and stimulating. The effect, however, is local, as the skin has very little if any absorbent power. The different nutritive and medicinal baths given with this end in view are practically worthless. Shipwrecked persons and those who are in the water a good portion of the time require less drink, not because the skin absorbs water, but because the long-continued contact of the water lessens the activity of the sweat glands, and consequently less water is thrown off from the body. When a daily bath is taken, unless the weather is very warm, soap is unnecessary, except to apply to those portions of the body which are exposed and where surfaces are in contact with each other. Too much soap removes the oily substances and renders the skin dry and harsh. Soldiers and others who are moving about and can not enjoy a regular bath should, if possible, bathe the portions of the body just referred to.

Special care should be given to the feet, not only as a matter of cleanliness, but to aid the proper locomotion. An ingrowing nail or neglected corn often renders a soldier unfit for duty. The feet should be bathed morning and evening; the nails frequently trimmed—not too close, as this causes the toes to become clubbed. Corns should be soaked in warm water and scraped, not cut.

The mouth and teeth should be cleaned twice daily with a soft brush; good soap may be used for this purpose. This procedure removes particles of food, etc., helps to preserve the teeth, and prevents oftentimes an unpleasant breath.

Dandruff is the scaly layer of the scalp; the excess can generally be removed by brushing; a fine comb should not be used. The scalp can occasionally be washed with warm water and soap or borax, and afterward washed with cold water. Too much soap or cosmetic changes the color of the hair. Inferior soap frequently irritates the skin, and should not be used.

Clothing.—It is generally accepted that woolen is the best material for clothing of all kinds. Its hygroscopic properties—i. e., the power which it possesses for absorbing water and its action as a nonconductor—make it particularly valuable for the above-named purpose. As a nonconductor it does not transmit the sun's rays in warm weather nor the body heat in winter. The latter heat is in a great measure regulated by the perspiration, which is readily absorbed by wool, and helps to cool the surface during a high temperature. Cotton and linen, being good conductors, transmit the sun's rays and also the warmth of the body. The chilly and clammy feeling which is often experienced when cotton is worn next the skin does not occur when the material is woolen. For this reason the latter is particularly indicated for wear by those who are subject to chest affections, rheumatism, etc. When it is necessary that the surface of the body should be kept uniformly warm, woolen should be worn at night as well as during the day. A common prejudice exists against the use of woolen as underwear, the objection being that it irritates the skin. At the present day woolen undergarments may be procured almost as soft as silk, and very thin. Silk as an article of clothing stands next to wool in value, linen and cotton last.

More or less cotton is generally found in woolen goods. If the amount is small, it does not materially interfere with the properties of the wool, and helps to prevent the excessive shrinking which would otherwise take place. The latter may be in a measure controlled if the garments are soaked and stirred in hot soap-suds and transferred to cold water (to remove soap) without wringing.

Woolen stockings are often objectionable, because the leather foot-wear, being also a bad conductor, keeps the feet tender, moist, and offensive. The popular notion that red woolen flannel is superior to the undyed is fallacious. This material is disagreeable, as the dye increases the absorption of offensive odors.

Head-wear should be worn as little as possible. It should be easy fitting, light in weight, and ventilated. Baldness frequently follows a disregard of these rules.

Comforters and other articles worn about the neck, except on special occasions, are to be condemned; they keep the skin moist and tender.

A number of coverings of clothing is warmer than one of equal thickness, due to the air between the several garments.

Leather clothing is a bad conductor, and very warm, but should only be worn in extremely cold, *dry*, or windy weather.

Rubber clothing should only be worn in wet weather. It prevents the proper evaporation from the skin, and keeps the body and underclothing moist.

Shoes should be made with broad soles, not too thick, to prevent the proper bending of the feet. Boots or shoes should never be worn on alternate feet. Boots sweat the legs, and should be worn only when necessary.

Parke's water-proof dressing is a very valuable preparation, and is made as follows : Dissolve carefully with heat half a pound of shoemaker's dubbing in half a pint of linseed-oil and half a pint of solution of India-rubber, and use on boots and shoes every two or three months.

Tight lacing, garters, and straps about the waist should be avoided ; the latter is said to favor the formation of hernia or rupture.

Food.—The function of food has been likened to the fuel which supplies the engine; but, aside from supplying material which keeps the different functions of the body in activity, it furnishes material for new tissue to supply the waste which is constantly going on. Food may be divided into organic and inorganic substances.

An organic substance is one which supplies the body with nourishment, heat, and motion, and forms new tissue to replace that which is worn out. In performing this it loses its identity, and when eliminated from the body is in the form of waste tissue.

Organic substances are divided into organic nitrogenized and organic non-nitrogenized. Organic nitrogenized matter or albuminous material contain nitrogen, and is the most important of all foods. It rebuilds worn-out tissues, supplies motion, strength, and nourishment. The white of egg, caseine of milk, and gluten of flour are examples of this class.

The organic non-nitrogenized substances do not furnish

material for such work as the nitrogenized, but supply heat as the result of oxidation, protect the deeper structure, give symmetry to the body, and are also used for food. Examples of this class are the starch and sugars (carbohydrates) and the fats (hydrocarbons).

It will thus be understood why an athlete chooses a diet composed principally of organic nitrogenized food, which forms muscles without materially increasing the body weight.

Inorganic substances are those which are taken into the system, perform a definite function, and are then removed from the body unchanged. Water and salt are examples of this class. Water exists in every tissue of the body, and is essential to their integrity. In cholera, where the water is rapidly removed, the body has a shrunken and pinched appearance, the features in some cases being unrecognizable. Salt is necessary to the proper interchange of nutrition. When deprived of this substance the body rapidly suffers.

All tissues of the body are constantly undergoing physiological change, and for this the food we eat must contain the different materials above enumerated. This is imperative, and has been confirmed by numerous experiments.

Ordinary articles of food contain 50 to 60 per cent of water. It is generally accepted that a diet should contain one part of nitrogen (contained in organic nitrogenized matter) and twelve to fifteen parts of carbon (found in organic non-nitrogenized food). Parke states that a man in good health and doing a good day's work must have about two thirds of an ounce of nitrogen and eight to twelve ounces of carbon.

The following interesting facts are given by Dr. Wilson: One ounce of albuminous material (organic nitrogenized) contains 70 grains of nitrogen and 212 grains of carbon; one ounce of the fat (hydrocarbon) contains 336 grains of carbon; one ounce of the starch and sugars (carbohydrates) contains 190 grains of carbon.

HYGIENE. 213

With these facts in view, a proper diet can be easily selected from the following table of Dr. Letheby, which includes the ordinary articles of food (uncooked); 20 per cent should be deducted for bone, and the same percentage for cooking:

	GRAIN PER POUND.			GRAIN PER POUND.	
	Carbon.	Nitrogen.		Carbon.	Nitrogen.
Split peas	2,699	248	New milk	599	44
Indian meal	3,016	120	Skim cheese	1,949	483
Barley meal	2,563	68	Chedder cheese	3,344	306
Rye	2,693	86	Bullock's liver	934	204
"Seconds" flour	2,700	116	Mutton	1,900	189
Oat meal	2,831	136	Beef	1,854	184
Baker's bread	1,975	88	Fat pork	4,113	106
Pearl barley	2,660	91	Dry bacon	5,987	95
Rice	2,732	68	Green bacon	5,426	76
Potatoes	769	22	Whitefish	871	195
Turnips	263	13	Red herring	1,435	217
Green vegetables	420	14	Dripping	5,456	...
Carrots	508	14	Suet	4,710	...
Parsnips	554	12	Lard	4,819	...
Sugar	2,955	...	Salt butter	4,585	...
Treacle	2,395	...	Fresh butter	6,456	...
Buttermilk	387	44	Cocoa	3,944	140
Whey	154	13	Beer and porter	274	1
Skimmed milk	438	43			

The following table is given by Dr. Dalton. A man in good health, taking active exercise in the open air and restricted to a diet of bread, butter, and fresh meat, with water and coffee for drink, consumes in one day:

Meat, 16 ounces.
Bread, 19 ounces.
Butter or fat, 3·5 ounces.
Water, 54 ounces.

Climate has considerable to do with the formation of the quantity and quality of the dietary. This is well marked in the arctic regions, where an enormous quantity of fat is consumed, thus furnishing, besides food, heat and protection.

Table from " Treatise on Hygiene," Notta and Firth, 1896.

Article of Food.	In 100 Parts.				
	Water.	Proteids.	Fats.	Carbohydrates.	Salts.
Beef, medium	69·5	20·5	8·4	....	1·6
Mutton, medium	75·79	17·11	5·77	....	1·33
Bacon	15·	8·8	73·3	....	2·9
Poultry	74·	21·	3·8	....	1·2
Flour, wheat of average quality	15·	11·	2·	71·2	·8
Wheaten bread	40·	8·	1·5	49·2	1·3
Rice	10·	5·	·8	83·2	·5
Oatmeal	15·	12·6	5·6	63·	3·
Peas, dried	15·	22·	2·	53·	2·4
Potatoes	74·	2·	·16	21·	1·
Cabbage	85·5	5·	·50	7·8	1·2
Eggs	73·5	13·5	11·6	....	1·
Milk, cow's	86·9	4·2	3·7	4·5	·7
Butter	12·	2·	85·	....	1·

Milk contains the essential elements necessary to a varied diet, and is the sole nourishment during infancy and the principal article of diet during the first five or six years of life, and even later. Milk is capable of sustaining life and health indefinitely in an adult. The different forms of aliments, however, may after childhood be advantageously taken in a more condensed form, as milk contains about 86 per cent of water. The specific gravity of good milk is between 1,017 and 1,036. There should be 10 or 12 per cent by volume of cream. When used for infants under six months of age the milk should be diluted with 25 to 100 per cent of water, and the addition of a small amount of sugar and cream. Milk is highly absorbent, and rapidly becomes tainted in an impure atmosphere. The only safety in using suspicious milk is to boil it; this destroys all germs. Cow's milk used for infant feeding is now generally Pasteurized—i. e., subjected to a temperature of about 155° Fahr. for one half to three quarters of an hour.

Both brown bread and eggs represent a varied diet, and

are very nutritious. Fresh eggs, when held to the light, are more transparent in the center, and stale ones at the ends. In a solution of one part of salt and ten of water good eggs sink and bad ones float.

Of the meats, beef has the greatest nutritive value, and can be taken indefinitely without becoming unpalatable.

Peas and beans are exceedingly nourishing. White bread is also of great value as an article of diet, particularly so when combined with milk. Sour bread may be utilized for food by toasting, as heat volatilizes the acid.

Cheese is rich in nitrogen, but hard to keep, and decays rapidly. Beef extracts and juices are very inferior to meats. They may be used as nutrients and stimulants for a short time; they do not, however, supply tissue waste.

Potatoes are very rich in starch, contain a large amount of water, and a small amount of vegetable acids.

Fresh fruits and vegetables are of great value; they assist in regulating the assimilative process, and act as a stimulus to the gastro-intestinal apparatus. They contain a large amount of vegetable acids.

Scorbutus or "scurvy," an exceedingly disagreeable and oftentimes fatal disease, is directly attributed to the absence of potatoes and other vegetables and fresh fruit in the diet. It often affects sailors and others who are deprived of the above articles of food and are restricted to a diet of salt meats. This disease comes on slowly, with a feeling of general debility and mental apathy, with a pale yellowish tint of skin, and insomnia; the gums become soft, and bleed; the breath has a fetid odor. There are also pains in the legs which may simulate rheumatism, great prostration, hæmorrhage from different portions of the body, diarrhœa, and often some chest affection is present. The treatment, if it can properly be carried out, is quick and effective; it consists mainly in supplying the needed dietary articles, combined with tonics, etc. The articles of food which are used as a preventive as well as in the treatment of this disease are known as "antiscorbutics, and comprise fresh fruits, potatoes, onions, cranberries, pickled cabbage, lime and lemon juice, cheap light wines and beers; raw walrus meat is highly recommended. As potatoes contain a certain amount of vegetable acids, they are regarded as antiscorbutics. Sugar, raisins, currants, yellow mustard, cresses, dandelions, and all varieties of cactus are said to be good. Pemmican, a mixture of dried meat pulverized and mixed with fat, is generally found among the stores of an arctic explorer.

A "ration" represents the *daily* allowance of food for one person.

Water.—An adult requires about 70 to 100 ounces of water daily; 20 to 30 ounces of this is contained in the food.

Horses and cattle need about six gallons daily.

The water-supply is derived from rain-water, springs, rivers, lakes, wells, etc.

Springs are the outlets of underground water, and, as a rule, give a pure and sparkling stream, the character of which is determined by the composition of the bed from which it is collected. As the rain-water, which contains considerable carbonic acid, passes through the ground, more or less change takes place, dependent upon the quality of the strata beneath. Spring-water is frequently rich in the salts of lime and magnesia. The presence of these salts in large quantities make insoluble compounds with soap, and give to water which contains them the name of "hard water." Mineral springs are caused by the chemical change which takes place under the surface.

Rain-water, if collected in the country and upon a proper receiver, is pure, whereas, if collected in cities or towns is suspicious, as the air in the vicinity of these places contains large quantities of organic and other matter which is injurious. Rain-water for drinking purposes should not be collected upon the roofs of houses unless specially prepared; neither are the ordinary wooden cisterns the proper receptacles for rain-water used for drinking; for this purpose they should be made of stone, slate, or other similar substances. For immediate use or in emergencies rain-water may be collected upon pieces of canvas and other material capable of holding water. Rain-water contains a comparatively small amount of the lime and magnesia salts, and is consequently "soft water."

The value and safety of river-water for drinking purposes depends upon its proximity to cities and large towns and the sewerage therefrom, which is a source of contamination; however, considerable of this matter is destroyed by the oxidation combined with the constant motion which is present. The numerous fresh-water plants also cause the destruction of organic matter. River-water may be classed as a "soft water."

Lake-water, although apparently stagnant, contains agents which keep up a more or less constant motion. Among these are the numerous springs which are found in the bottom of the lakes. Considerable of the organic matter present is oxidized by the sunlight and air, and the tranquillity of the water allows a large amount of the organic matter to be precipitated. The vegetable growth in lakes also acts as purifiers, consequently water taken from a large lake should be pure; however, water from small lakes or ponds should not be used for drink.

Deep wells in the country and small towns constitute the usual means of securing drinking-water in these places.

Well-water contains considerable lime and magnesia, and is consequently more or less hard. While deep wells in the country and small towns represent probably the best means of securing a good and wholesome drinking-water, it must be remembered that they should not be within 100 feet of any privy, cesspool, or other means of pollution, which will soak through the earth for a long distance. Typhoid fever in the country is commonly contracted by drinking from a well in close contact with a privy-vault containing the discharge of persons suffering from this disease. Water from wells in cities and large towns should never be used for drinking purposes. Boiling or filtering through a Pasteur or Berkefeld filter is the only sure means by which suspicious water can be rendered safe for drinking. Boiling is simple and easily performed, and should be done whenever there is reason to believe that the water may be polluted, as during the prevalence of cholera and typhoid fever. Boiled water is more or less flat, resulting from the removal of the gas or air which it contains by the action of the heat. It may be aërated or livened up by agitating in an open churn, or pouring from one receptacle to another, thus entangling some air, which makes the water more palatable. Boiling usually renders water somewhat softer.

A simple means of filtering water is to tie a piece of double thickness canton flannel over the water tap, or over a receptacle, and allowing the water to pass through it. The filter should be frequently changed.

Turbid water can be cleaned by adding five or ten grains of alum to a gallon; then boil and filter. Other means for effecting this purpose are by pieces of cactus-leaves, or five or six grains of tannin to a gallon, exposure to sun and air, and filtering through sand.

Air.—Pure air is essential to health; the different tissues of the body demand an uninterrupted supply of oxygen for their maintenance; this is derived from the air, the composition of which is as follows:

In 100 parts—

Oxygen. $20\cdot96$; nitrogen, 78; argon, 1; carbonic acid, $\cdot04$.

Ammonia, ozone, watery vapor, organic matter, and mineral salts—traces.

With the exception of the latter group, the relative proportions of the different ingredients of air remain about the same everywhere. The impurities in the air are derived from filthy and badly ventilated rooms and buildings, apartments inhabited by unhealthy persons and those suffering from contagious diseases, offensive

HYGIENE. 219

trades, emanations from sewers, cesspools, marshes, mines, etc., and may be divided into suspended and gaseous matter. The former includes material from the animal and vegetable world, germs of disease, etc. Among the gaseous matter, carbonic acid is the most constant and prominent; a small amount of this gas is always present in what may be called pure air, and represents about ·04 part per 100. It is the increased amount of this gas, associated with organic matter, watery vapor, etc., which is responsible for the uncomfortable sensation experienced in public gatherings, improperly ventilated schools, etc. Persons subjected to continued defective ventilation suffer from depression, headache, gastric disturbance, etc., and are particularly susceptible to disease. The increased amount of carbonic acid in the air in large assemblages is derived mainly from the expired air of those present; about four per cent of the oxygen taken in with each inspiration is appropriated by the blood passing through the lungs. An equal amount (four per cent) of carbonic acid, representing effete or worn-out matter, is thrown off in the expired air, and is unfit for respiration; in addition to the carbonic acid, a varying amount of watery vapor and organic matter is also exhaled.

In an overcrowded room with improper ventilation the oxygen becomes rapidly diminished and replaced by carbonic acid; this change may not be noticed until some time has elapsed; the air of the room becomes offensive, and those present feel sleepy, with more or less headache, and embarrassment of respiration. Among the more terrible results of defective ventilation may be mentioned the tragedy that occurred in Calcutta about the middle of the last century, when one hundred and forty-six English prisoners were thrown into the jail or "Black Hole" of Calcutta. This consisted of a room about eighteen feet square, with two small windows covered with iron bars. The prisoners were put in this apartment in the latter part of the day or evening, and when the door was opened in the morning all but twenty-three were found dead. Some of the survivors subsequently died as the result of this confinement.

The fresh air necessary to ventilate an apartment should be enough to destroy all sensible odors or impurities which would result from imperfect ventilation, so that a person coming from the outside should from the odor appreciate no difference in the air.

Apartments occupied by the sick always require relatively more fresh air, as there is an increased exhalation from the skin and lungs containing a larger amount of organic matter, etc.

An increased amount of fresh air is also needed where artificial light is used (excepting electricity), as this method of illumination consumes a good amount of oxygen. Contamination of the air may also occur where coal is used for heating purposes, particularly in defective stoves. Fireplaces constitute a most excellent means of ventilating a room, the impure air being drawn up the chimney and replaced by fresh air which is admitted through cracks under doors, etc.

Electric light is far superior to every other form of artificial light, inasmuch as it consumes no oxygen, and gives off no product of combustion by which the air may become polluted; it also gives off a very small amount of heat.

At least four or five hundred cubic feet of air space should be allowed for every occupant of an apartment. It may not be out of place in this connection to speak of the importance of fresh air where cases of phthisis or tuberculosis (consumption) exist. It is now known that this disease is due to a special germ or bacillus which is present in large numbers in the expectoration or sputum; the latter becomes dried and diffused as dust in the air, and by being inhaled may infect other persons, particularly those who are in close and continued contact with the patient. In other words, consumption is due to the reception of the tubercle bacillus communicated from one infected with the disease to another, and can be prevented if the proper precautions are taken (see DISINFECTION).

Exercise.—Organs and muscles which do not receive the proper exercise become smaller, or undergo what is known as "atrophy." Therefore exercise of all portions of the body is necessary. Insufficient exercise favors the accumulation of carbon in the system, with the consequent bad results. Those who follow sedentary habits and are confined at desks, etc., are prone to chest troubles from diminished lung capacity.

Work in a gymnasium, etc., does not represent the exercise essential to health. It should be in the open air, and not too violent. Horseback and bicycle riding are good types of proper exercise combined with pleasure, the latter being an important consideration. In the selection of a form of exercise, walking may be considered as the common means by which the different portions of the body can be properly stimulated. It has been estimated that a walk of ten miles distributed over the day is about the proper amount of exercise which should be taken by a healthy adult in twenty-four hours. This is said to be equivalent to lifting one hundred and fifty tons one foot from the ground daily. Too much or too violent exercise is injurious, particularly in the weak and delicate.

CHAPTER XX.

TRANSPORTATION OF THE WOUNDED.

It is an important part of the management of emergencies and accidents that means should be supplied to remove the patient as quickly and carefully as possible to a place where the proper treatment can be applied. Transportation may be effected by the use of a stretcher (litter), or by one or more persons carrying the one injured. The use of a litter is superior to the latter means, and should always be employed if practicable.

The following essential qualities for a stretcher are suggested by Longmore (Pilcher, "Transportation of the Wounded; Reference Handbook of Medical Science"):

"1. A support for the patient, firm and comfortable, but capable of being readily cleansed.

"2. Lightness, to facilitate carriage by bearers.

"3. Strength, to resist shock from rough usage.

"4. Simplicity of construction, combined with—

"5. Capability of being folded up to economize space in stowage and to lessen liability to injury.

"6. Such a connection of the component parts as to prevent risk of loss.

"7. Provision for keeping the patient a certain distance above the ground when the litter is laid down.

"8. Economy in cost."

The different forms of litters are as numerous as the varieties of splints, and are either manufactured or extemporized.

MANUFACTURED LITTERS.—The Halstead litter, which

is used in the U. S. Army, complies with the above essential points, and its construction is detailed by Captain Pilcher, Assistant Surgeon U. S. A., as follows : "The Halstead litter consists of two poles of seasoned white ash, eight feet long and one inch and a half square, the ends of which are rounded off for handles. These poles are connected by braces, one at either end, each consisting of two pieces of wrought-iron one inch wide and three eighths of an inch thick ; one piece is fifteen and the other twelve inches in length, hinged in the center of the litter ; the longer one overlapping the shorter three inches and a half, and when the litter is open shutting on a bolt or pin, forming a stiff shoulder for the hinge and preventing the stretcher from accidentally closing. The braces attached to the under side of the poles just external to the legs, so as to form a shoulder against which the legs impinge, are fastened to the side-poles with heavy screws, pieces of common hoop-iron being placed underneath them to prevent their wearing the wood. The legs, made of seasoned white ash like the poles, are fourteen inches and a half long, one inch thick, one inch and seven eighths wide at the top, and tapering to one inch and three eighths at the bottom ; they are fastened to the poles with screw-bolts having washers under the heads and rivets through the upper end of the legs to prevent their splitting. Over the top of this framework is tightly drawn a strip of canvas which is fastened on the outside of the poles with six-ounce tacks, forming a bed five feet and eleven inches in length, and twenty-three inches and a half in width, allowing the poles to project as handles for a distance of thirteen inches at one end and twelve inches at the other. The shoulder-straps weigh eight ounces, and are made of striped cotton webbing two inches and a half wide by fifty inches long, with a five-inch loop at one end and at the other end a leather strap twenty-two inches and a half long by one inch and one sixth wide, with a buckle to loop around the handles of the litter at any length desired. A hair-pillow covered

Fig. 86.—Marsh's stretcher.

with canvas also accompanies it. The litter complete weighs only twenty-three pounds and a quarter."

A very ingenious and practical litter has been devised by Major E. T. T. Marsh, Surgeon of the Seventy-first Regiment, N. G. S. N. Y. It consists of two ash side-bars, eight feet long and about two inches in diameter, attached to a piece of canvas six feet long and two feet wide. The side-bars are hinged (but not disconnected) in the middle, the joint being strengthened by a brass ferrule which slips over it. The two spreaders, one at each end, which keep the side-bars apart, are made of wood. One end of each spreader is attached by a hinge to a side-bar, the other end is capped with a metal crutch to fit the opposite side-bar. The stretcher, which weighs only ten pounds, can be folded in a very compact form (Fig. 86).

The above are given as examples of manufactured litters, although numerous others are in general use.

EXTEMPORIZED LITTERS.—The extemporized litters include any agent (not manufactured for the purpose) which can be utilized to carry the sick and wounded. Among the articles which may be employed are window-shutters, boards, doors, bed-frames, mattresses, benches, tables, chairs, blankets—the last may be carried by two or more persons, or the four corners may be tied to two poles; rugs, hammocks, carpets, may be used in the same manner. It is said that during an Indian campaign General Jackson carried the wounded on litters made of the skin of oxen strung between two guns or poles. Guns or poles may be passed through sacks (openings being made at the bottom), or coats (Fig. 87), or trousers. Poles may be retained at the proper distance apart by rope or telegraph-wire.

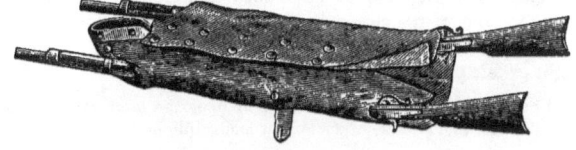

Fig. 87.—Extemporized stretcher.

DRILL REGULATIONS FOR THE HOSPITAL CORPS, U. S. ARMY.

April 30, 1896.

BY AUTHORITY OF THE SURGEON GENERAL.

1. The senior medical officer of the detachment is responsible for the theoretical and practical instruction of the officers, non-commissioned officers, and privates. He requires them to study and recite these regulations so that they can explain thoroughly every movement before it is put into execution.

THE DETACHMENT.

2. The detachment, when formed, is in single rank, privates of the Hospital Corps on the right, company bearers on the left, each class graduated in size, the tallest men on the right.

POSTS OF OFFICERS AND NON-COMMISSIONED OFFICERS.*

3. The medical officer commanding is three paces in front of the center of the detachment, the junior medical officers according to rank from right to left are in the line of file closers two paces in rear. Medical officers above the rank of lieutenant are four paces in rear of the file closers.

Medical officers take posts in their respective lines at equal intervals; if only one, he is opposite the center; if two, one is opposite the center of each half of the detachment; if three, one is opposite the center, the other as with two.

The senior non-commissioned officer is two paces in rear of the second file from the right, on the right of the line of file closers.

The second non-commissioned officer is on the right of the rank and is right guide of the detachment.

The third non-commissioned officer is on the left of the rank and is the left guide.

The remaining non-commissioned officers are distributed along the line of file closers from right to left, according to rank.

If necessary, a suitable private may be designated to act as right or left guide.

* For the purposes of these drill regulations the term non-commissioned officer includes hospital stewards and acting hospital stewards.

TO FORM THE DETACHMENT.

4. At the signal for the *assembly*,* the senior non-commissioned officer takes his position in front of where the center of the detachment is to be, and facing it, commands:

Fall in.

The second non-commissioned officer, or a designated private, places himself facing to the front, where the right of the detachment is to rest, and at such a point that its center will be six paces from and opposite the senior non-commissioned officer; he closes his left hand and places the knuckles against the waist above the hip, wrist straight, back of hand to the front. The men, with left arm in same position, assemble rapidly at attention, so that the right arm of each man rests lightly against the left elbow of the man next on his right, each dropping the left hand as soon as the man next on his left has his interval.

The other non-commissioned officers then take their posts.

The senior non-commissioned officer calls the roll, each man answering, "Here," as his name is called.

TO SIZE THE DETACHMENT.

5. The men being in line as described, the senior non-commissioned officer faces them to the right and arranges them according to height, tallest man in front; he then faces them to the left into line. The detachment being sized, habitually forms in the same order.

6. The senior non-commissioned officer commands:

1. *Count*, 2. Fours.

Beginning on the right the men count *One, Two, Three, Four,* and so on to the left. The guides do not count.

If the four on the left consists of less than three men they are ordinarily assigned to other fours and placed in the line of file closers, each in rear of the four to which assigned. He then commands:

1. *Count*, 2. Squads;

when each No. 1 calls out the number of his squad in numerical order from right to left.

---

* The *assembly* may be sounded by bugle or whistle.

The officer commanding having approached the front and center of the detachment, the senior non-commissioned officer faces about, salutes* him, reports the result of the roll call, and then, without command, takes his post, passing around the right flank.

The junior medical officers take their posts as soon as the non-commissioned officer has reported:

ALIGNMENTS.

7. The officer commanding, having received the detachment, commands:

1. *Right* (or *left*), 2. Dress, 3. Front.

At the command *dress*, the men place the left hand above the hip, turn the head and eyes in the direction of the guide, and dress up to the line; the officer commanding verifies the alignment. At the command *front*, the men turn the head and eyes to the front and drop the left hand.

In all alignments, excepting of the file closers, the left hand is placed above the hip, and at *front* dropped to the side. The detachment is aligned whenever necessary.

MARCHINGS.

8. When the execution of a movement is improperly begun and the instructor wishes to begin it anew for the purpose of correcting it, he commands: *As you were;* at which the movement ceases and the former position is resumed.

9. The length of the full step in quick time is 30 inches measured from heel to heel, and the cadence is at the rate of 120 steps per minute.

TO MARCH IN LINE.

10. Being in line at a halt:

1. *Forward*, 2. *Guide right* (or *left*), 3. March.

The men step off, the guide marching straight to the front.

---

* The senior non-commissioned officer when armed with the saber salutes by bringing it to the first position of *inspection arms* [Par. 176], and then to the *carry*. When not armed he raises the right hand smartly till the forefinger touches the lower part of the headdress above the right eye, thumb and fingers extended and joined, palm to the left, forearm inclined at about forty-five degrees, hand and wrist straight; and then drops

The instructor sees that the men preserve the interval and alignment.

To change the guide: *Guide left* (or *right*).

11. If the men lose step, the instructor commands: STEP.

The men glance toward the side of the guide, retake the step, and cast their eyes to the front.

TO MARCH BACKWARD.

12. Being at a halt:

   1. *Backward*, 2. *Guide right* (or *left*), 3. MARCH.

At the command *march*, step back with the left foot 15 inches straight to the rear, measuring from heel to heel, then with the right, and so on, the feet alternating.

TO MARCH TO THE REAR.

13. Being in march:

   1. *To the rear*, 2. MARCH, 3. *Guide right* (or *left*).

At the command *march*, given as the right foot strikes the ground, advance and plant the left foot; then turn on the balls of both feet, face to the right about, and immediately step off with the left foot.

If marching in double time, turn to the right about, taking four short steps in place, keeping the cadence, and then step off with the left foot.

TO SIDE STEP.

14. Being at a halt:

   1. *Right* (or *left*) *step*, 2. MARCH.

At the command *march*, carry the right foot 12 inches to the right, keeping knees straight and shoulders square to the front; as soon as the right foot is planted, bring the left foot to the side of it, and continue the movement.

TO MARCH BY THE FLANK, IN COLUMN OF FILES.

15. Being in line at a halt:

   1. *Right* (or *left*), 2. FACE, 3. *Forward*, 4. MARCH.

the arm quietly by the side. This is the salute for all enlisted men without arms.

Being in march:

    1. *By the right* (or *left*) *flank*, 2. MARCH.

At the command *march*, given as the right foot strikes the ground, advance and plant the left foot, then face to the right in marching and step off in the new direction with the right foot.

To halt the column of files: 1. *Detachment*, 2. HALT; and to face it to the front: 3. *Left* (or *right*), 4. FACE.

MARCHING IN COLUMN OF FILES, TO MARCH IN LINE.

16. 1. *By the left* (or *right*) *flank*, 2. MARCH, 3. *Guide right* (or *left*).

TO CHANGE DIRECTION IN COLUMN OF FILES.

17. Being in march:

1. *Column right* (or *left*); or, 1. *Column half right* (or *half left*); 2. MARCH.

The leading file wheels to the right. The other files follow the first and wheel on the same ground.

Being at a halt:

    1. *Forward*, 2. *Column right* (or *left*), 3. MARCH;

or:

    2. *Column half right* (or *half left*), 3. MARCH.

THE OBLIQUE MARCH.

18. Being in line at a halt, or in march:

    1. *Right* (or *left*) *oblique*, 2. MARCH.

At the command *march*, each man half faces to the right, at the same time stepping off in the new direction. He preserves his relative position, keeping his shoulders parallel to those of the man

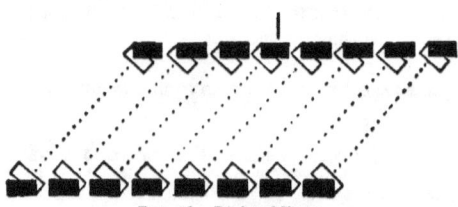

PAR. 18.—Right oblique.

next on his right, and so regulates his step as to make the head of this man conceal the heads of the other men in the rank; the rank remains parallel to its original front.

At the command *halt*, the men halt, faced to the front.

To resume the original direction:

1. *Forward*, 2. MARCH.

The men half face to the left in marching and then move straight to the front.

TO MARCH IN DOUBLE TIME.

19. The length of the full step in double time is 36 inches; the cadence is at the rate of 180 steps per minute.

Being in line at a halt:

1. *Forward*, 2. *Guide right* (or *left*), 3. *Double time*, 4. MARCH.

At the third command the hands are raised until the forearms are horizontal, fingers closed and toward the body, the elbows to the rear.

TO PASS FROM QUICK TO DOUBLE TIME, AND THE REVERSE.

20. 1. *Double time*, 2. MARCH.

At the command *march*, given as the left foot strikes the ground, advance the right foot in quick time, and step off with the left foot in double time.

To resume quick time:

1. *Quick time*, 2. MARCH.

At the command *march*, given as either foot is coming to the ground, the detachment resumes quick time.

TURNINGS.

TO TURN AND HALT.

21. Marching in line:

1. *Detachment right* (or *left*),
2. MARCH, 3. FRONT.

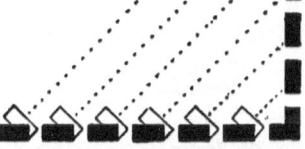

PAR. 21.—Detachment right.

At the command *march*, the right guide halts and faces to the right; the other files half face to the right in marching, and without changing the length or cadence of the step place themselves successively upon the alignment established by the right guide; all dress to the right without command. The instructor verifies the alignment from the pivot flank and commands: FRONT.

If at a halt, the movement is executed in the same manner.

Detachment half right (or *half left*) is executed in the same manner, except that the guide makes a half face to the right.

TO TURN AND ADVANCE.

22. Marching in line:

1. *Right* (or *left*) *turn*, 2. MARCH, 3. *Forward*, 4. MARCH, 5. *Guide right* (or *left*).

At the second command, the guide marches by the right flank, taking the short step without changing the cadence; the other men half face to the right in marching, and moving by the shortest line successively place themselves on the new line, when they take the short step (15 inches).

When the last man has arrived on the new line, the fourth command is given, when all resume the full step.

During the turn the guide is, without command, on the pivot flank. The guide is announced on resuming the full step.

If at a halt, the movement is similarly executed, and in quick time, unless the command *double time* is given.

Right (or *left*) *half turn* is executed in the same manner, except that the guide makes a half face to the right.

Should the command *halt* be given during the execution of the movement, those men who are on the new line halt; the others halt on arriving on the line; all dress to the right without command.

The instructor verifies the alignment from the pivot flank and commands: FRONT.

MARCHING IN LINE, TO EFFECT A SLIGHT CHANGE OF DIRECTION.

23. *Incline to the right* (or *left*).

Each man advances the left shoulder and marches in the new direction.

TRANSPORTATION OF THE WOUNDED. 233

BEING IN LINE TO MARCH BY THE FLANK, IN COLUMN OF FOURS.

24. 1. *Fours right* (or *left*), 2. MARCH.

Each four wheels ninety degrees to the right on a fixed pivot, the pivot man turning strictly in his place; the man on the marching flank maintains the full step, moving on the arc of a circle with the pivot man as the center; the men dress on the marching flank, shorten their steps according to their distance from it, and keep their intervals from the pivot. Upon the completion of the

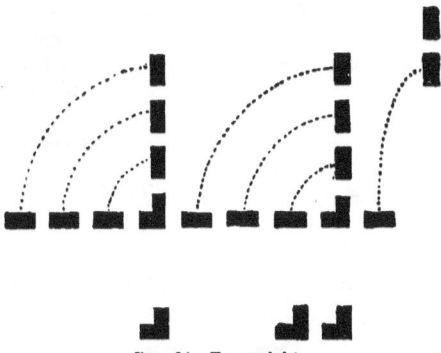

PAR. 24.—Fours right.

wheel, each four takes the full step, marching in a direction parallel to its former front; the second four is one hundred inches from the first four, and so on to the rear of the column; the right and left guides place themselves forty-four inches in front and rear respectively of the left file of the leading and rear fours; the file closers face to the right and maintain their relative positions.

The officer commanding in columns of fours, twos, and files is by the side of the leading guide on the flank opposite the file closers; he takes this position at the command *march*.

The leading and rear guides in columns of fours, twos, and files are in front and rear respectively of the leading or rear file on the side opposite the file closers.

The file closers march two paces from the flank of the column and see that the fours maintain their distances.

In all changes by fours from line into column and column into line, or from column of fours into twos, files or the reverse, and in all wheels about by fours, either in line or column, the officer commanding and guides take their proper places in the most convenient way as soon as practicable.

All wheels by fours, except in changing direction, are executed on a fixed pivot.

These rules are general.

BEING IN LINE TO FORM COLUMN OF FOURS AND HALT.

25. 1. *Fours right* (or *left*), 2. MARCH, 3. *Detachment*, 4. HALT.

The command *halt* is given as the wheel is completed; all dress toward the marching flank.

In column of fours, the ranks dress toward the side of the guide.

In all wheelings by fours, the forward march is taken upon the completion of the movement, unless the command *halt* be given.

At the command *halt*, given as either foot is coming to the ground, the foot in rear is brought up and planted without shock by the side of the other.

MARCHING IN COLUMN OF FOURS, TO CHANGE DIRECTION.

26. 1. *Column right* (or *left*), 2. MARCH.

The leading four wheels on a movable pivot; the pivot man takes steps of ten inches in quick time, and twelve inches in double time, gaining ground forward so as to clear the wheeling point; the wheel completed, the full step is taken; the man on the side of the guide follows forty-four inches in rear of the guide; the other fours move forward and wheel on the same ground. If the change of direction be toward the side of the guide, he shortens his step and wheels as if on the pivot flank of a rank of four; if the change be to the side opposite the guide, he wheels as if on the marching flank of a rank of four.

Column half right (or *half left*) is similarly executed, each four wheeling forty-five degrees.

TO PUT THE COLUMN OF FOURS IN MARCH AND CHANGE DIRECTION AT THE SAME TIME.

27. 1. *Forward*, 2. *Column right* (or *left*); or, 2. *Column half right* (or *half left*), 3. MARCH.

TRANSPORTATION OF THE WOUNDED. 235

BEING IN LINE, TO FORM COLUMN OF FOURS AND CHANGE DIRECTION.

28. 1. *Fours right* (or *left*), 2. *Column right* (or *left*); or, 2. *Column half right* (or *half left*), 3. MARCH.

BEING IN LINE, TO MARCH IN COLUMN OF FOURS TO THE FRONT.

29. 1. *Right* (or *left*) *forward*, 2. *Fours right* (or *left*), 3. MARCH.

At the command *march*, the right guide places himself in front of the left file of the right four; the right four moves straight to the front, shortening the first three or four steps; the other fours wheel to the right, each on a fixed pivot; the second four when its wheel is two-thirds completed, wheels to the left on a movable pivot and follows the first; the other fours having wheeled to the right, move forward, and each wheels to the left on a movable pivot, so as to follow the second.

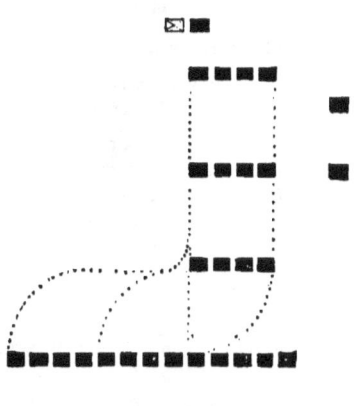

PAR. 29.—Right forward, fours right.

BEING IN COLUMN OF FOURS, TO CHANGE THE FILE CLOSERS FROM ONE FLANK OF THE COLUMN TO THE OTHER.

30. 1. *File closers on left* (or *right flank*), 2. MARCH.

At the first command the file closers close in to the flank of the column, and at the command *march*, dart through the column.

TO OBLIQUE IN COLUMN OF FOURS, AND TO RESUME THE DIRECT MARCH.

31. 1. *Right* (or *left*) *oblique*, 2. MARCH.

Each four obliques as prescribed (Par. 18).

To resume the direct march:

1. *Forward*, 2. MARCH.

TO MARCH IN COLUMN OF FOURS TO THE REAR.

32. 1. *Fours right* (or *left*) *about*, 2. MARCH.

Each four wheels 180 degrees to the right.

The file closers do not pass through the column, but gain the space to the right or left necessary to preserve their interval from the flank.

TO FORM LINE FROM COLUMN OF FOURS.

33. To the right or left:

1. *Fours right* (or *left*), 2. MARCH, 3. *Guide right* (or *left*); or, 3. *Detachment*, 4. HALT.

At the command *march*, the fours wheel to the right.

The guide is announced, or the command *halt* is given the instant the fours unite in line.

If the line be formed toward the side of the file closers, they close in to the flank of the column at the first command, and at the command *march* dart through the column.

34. On right or left:

1. *On right* (or *left*) *into line*, 2. MARCH, 3. *Detachment*, 4. HALT, 5. FRONT.

At the command *march*, the leading four wheels to the right on a movable pivot and moves forward, dressing to the right; the guide places himself on its right; each of the other fours marches a distance equal to its front beyond the wheeling point of the four next preceding, wheels to the right and advances as explained for the first four; the rear guide places himself on the left of the rear four after it halts.

At the command *halt*, given when the leading four has advanced a suitable distance in the new direction, it halts and dresses to the right; the other fours successively halt and dress upon arriving in line.

The command *front* is given when the left four completes its dressing.

If the movement be executed toward the side opposite the file closers, each follows the four nearest him, passing in front of the following four.

TRANSPORTATION OF THE WOUNDED.

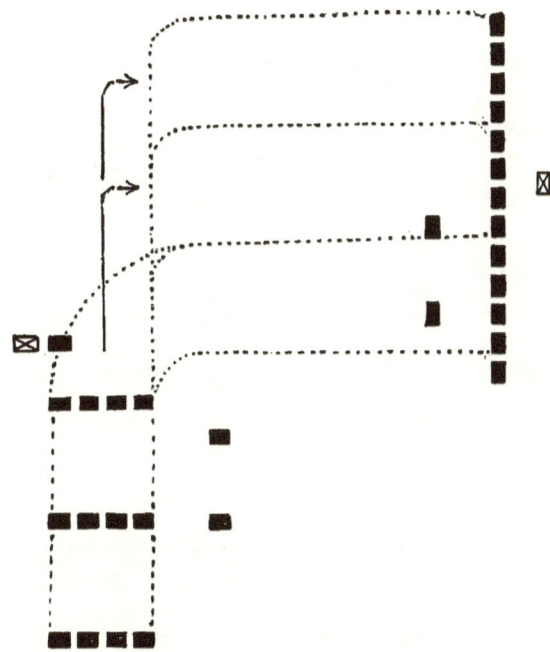

PAR. 34.—On right into line.

35. To the front:

1. *Right* (or *left*) *front into line*, 2. MARCH, 3. *Detachment*, 4. HALT, 5. FRONT.

At the command *march*, the leading four moves straight to the front, dressing to the left; the guide in front places himself on its left; the other fours oblique to the right till opposite their places in line, when each marches to the front.

At the command *halt*, given when the leading four has advanced a suitable distance, it halts and dresses to the left; the other fours halt and dress to the left upon arriving in line; the guide in rear places himself on the right of the rank upon the

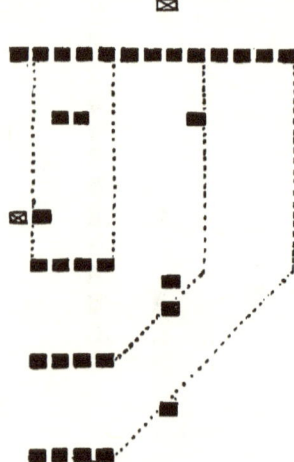

PAR. 35.—Right front into line.

arrival of the last four in line; the commnad *front* is given when the last four completes its dressing.

If the movement be made toward the side of the file closers, they dart through the column as the oblique commences.

If marching in double time, or in quick time, and the command be *double time*, the command *guide left* is given immediately after the command *march;* the leading four moves to the front in quick time; the other fours oblique in double time, each taking the quick time and dressing to the left upon arriving in line.

BEING IN LINE, TO FACE TO THE REAR AND TO MARCH TO THE REAR.

36. 1. *Fours right* (or *left*) *about,* 2. MARCH, 3. *Detachment,* 4. HALT; or, 3. *Guide right* (or *left*).

The fours wheel about; the file closers dart through the nearest intervals.

37. The detachment at a halt, may be marched a few paces to the rear by the commands:

1. *Detachment,* 2. *About,* 3. FACE, 4. *Forward,* 5. *Guide right* (or *left*), 6. MARCH;

or, if in march, by the commands:

1. *To the rear,* 2. MARCH, 3. *Guide right* (or *left*).

The file closers on facing about maintain their relative positions.

BEING IN COLUMN OF FOURS, TO FORM COLUMN OF TWOS.

38. This movement is always executed toward the file closers; it is used only for the purpose of reducing the front of the column

to enable it to pass a defile or other narrow place, immediately after which the column of fours should be re-formed.

1. *Right* (or *left*) *by twos*, 2. MARCH.

At the command *march*, the two files on the right of each four move forward; the two files on the left take the short step till disengaged, when they oblique to the right and follow the right files.

The distance between ranks in column of twos is forty-four inches; the guides take the same distance in front and rear of the column.

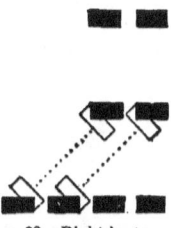

PAR. 38.—Right by twos.

BEING IN COLUMN OF FOURS AT A HALT, TO FORM COLUMN OF FILES.

39. 1. *Right* (or *left*) *by file*, 2. MARCH.

At the command *march*, the right file of each four moves forward, followed in succession by the files on his left, who oblique to the right, the men keeping closed as nearly as possible to facing distance; the guides take the same distance.

If marching, the leading file continues the march; the others take the short step, oblique to the right and follow the leading file.

Column of files from column of twos is similarly executed.

40. A column of twos or files changes direction, is halted and put in march by the same commands as a column of fours.

The march in column of fours, twos, or files is always in quick time unless otherwise ordered.

41. A column of fours, twos, or files may be faced to the rear, or to a flank, and marched a short distance, but no other movements should be executed until the column is again faced to the original front. The officers and non-commissioned officers face with the column and retain their positions.

BEING IN COLUMN OF TWOS OR FILES, TO FORM COLUMN OF FOURS.

42. These movements are always executed away from the file closers.

Marching in column of twos:

1. *Form fours*, 2. *Left* (or *right*) *oblique*, 3. MARCH.

At the command *march*, the leading two of each four take the short step; the rear two oblique to the left until they uncover the leading two, when they move to the front, and the fours having united, all resume the full step.

Being in column of files:

43. 1. *Form fours*, 2. *Left* (or *right*) *oblique*, 3. MARCH.

At the command *march*, the leading file of each four halts; the other files oblique to the left, and place themselves successively on the left of the leading file, the distance between the fours being one hundred inches.

PAR. 42.—Form fours, left oblique.

Column of twos is formed from column of files on the same principles.

THE RESTS.

44. Being at a halt, to rest the men:

FALL OUT; or REST; or AT EASE.

At the command *fall out*, the men may leave the ranks, but will remain in the immediate vicinity.

At the command *fall in*, they resume their former places.

At the command *rest*, the men keep one heel in place, but are not required to preserve silence or immobility.

At the command *at ease*, the men keep one heel in place and preserve silence, but not immobility.

45. To resume the attention:

1. *Detachment*, 2. ATTENTION.

The men take the position of the soldier and fix their attention.

46. 1. *Parade*, 2. REST.

Carry the right foot six inches straight to the rear, left knee slightly bent; clasp the hands in front of the center of the body, left hand uppermost, left thumb clasped by thumb and forefinger of right hand; preserve silence and steadiness of position.

To resume the position of the soldier:

1. *Detachment*. 2. ATTENTION.

TRANSPORTATION OF THE WOUNDED.

TO DISMISS THE DETACHMENT.

47. Being in line at a halt, the officer commanding directs the senior non-commissioned officer: *Dismiss the detachment.* The officers fall out, the senior non-commissioned officer salutes, steps in front of the detachment, and commands: DISMISSED.

LITTER DRILL.

THE LITTER AND SLING.

48. The regulation hand litter consists of a canvas bed, 6 feet long and 22 inches wide, made fast to two poles 7½ feet long, and stretched by two jointed braces. The ends of the poles form the handles, 9 inches long, by which the litter is carried. The fixed iron legs are stirrup-shaped, 4 inches high and 1¾ inches wide. A strap is permanently fastened to the bottom of each pole near the legs. When the litter is open each strap lies under the pole to which it is attached, its free end buttoned to a stud; when the litter is strapped it is passed over the canvas and the free end buttoned to a stud under the opposite pole.

49. One regulation sling is issued to each private as part of his equipment. It is made of blue webbing, 2¼ inches wide, with a leather-lined loop at each end, and a slide to regulate its length.

Before falling in, each man verifies the length of his sling so that it is always exactly adjusted to his size, places it smoothly over his shoulders, the slide on the right side, passing the loops under the belt, and, during drills with the litter, never removes it from the body. The loops, when off the handles, are invariably secured under the belt.

50. When the detachment is formed for drill or instruction, officers do not wear swords. The instructor will require that the clothing of the men be clean and neatly adjusted; that the privates of the Hospital Corps fall in equipped with pouch, belt, and sling, and that the company bearers wear belts and brassards.* Hospital stewards and acting hospital stewards wear the belt, but no equipment of any kind, unless specially ordered.

---

* When practicable, it is advisable to have the men who are to represent the wounded dressed in fatigue suits; they remain in the line of file closers until needed.

51. For purposes of litter drill each set of fours is a litter squad. The litter squad is marched by the commands applicable to a set of four, substituting "litter" for "four."

52. No. 1 is the squad leader; he commands his squad and is responsible for it; in his absence, No. 4, and both Nos. 1 and 4 being absent, No. 3 commands.

With reduced numbers, No. 1 ordinarily assumes the duties of No. 3, and No. 4 of No. 2. No. 1 being absent, No. 4 assumes his duties and *vice versa*.

53. The instructor will make such changes in the *personnel* of the sets of four as he deems advisable. The selection of No. 1 should be determined by the intelligence and experience of the men; No. 4 should be as near in size as possible to No. 1, and No. 2 to No. 3. The fours are then counted again if necessary.

54. A litter is said to be *strapped* when folded, the canvas doubled smoothly on top and secured by the straps. It is said to be *closed* when folded and unstrapped, the free ends of the straps buttoned to the studs on their respective poles.

55. The foot, or front, of a *grounded* or *open* (unloaded) litter is the end farthest from the approaching squad, unless otherwise designated. The foot of a loaded litter is always the end corresponding to the feet of the patient.

MANUAL OF THE LITTER.

Having assigned the medical officers and the non-commissioned officers to appropriate duties, the instructor commands:

56. 1. *Procure litter*, 2. *Right* (or *left*) *face*, 3. MARCH.

At the first command the Nos. 3 step one pace to the front, at the second command they face as required, and at the third proceed in column of files, by the nearest route, to the (strapped or closed) litters. They each take one, placing it on the right shoulder at a slope of at least 45 degrees, canvas down (par. 59), and promptly return in the reverse order, each man resuming his place by passing through his interval one pace to the rear, facing about and stepping into line.

If the litters are in front of the detachment, the Nos. 3 may be marched directly forward, converging toward them, and then back, diverging to their intervals.

This march should be supervised by a non-commissioned officer. It can be executed in double time.

With but one squad the commands are simply *procure* (or *return*) *litter*, MARCH; when the bearer proceeds and returns by the shortest practicable route.

57. In all motions from the *shoulder*, or to the *shoulder*, the litter should invariably be brought to the *vertical position* against the shoulder, one pole in front of the other, canvas to the left, both hands grasping the front pole, the left above the right, and the left forearm horizontal.

This position should be taken by the bearer when passing through his interval to resume his place in the line (par. 56), and in any formation or movement in which there may be danger of the lower or upper handles of the litter striking neighboring men, after which the *shoulder* is resumed without command.

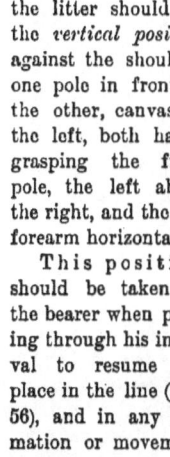

PAR. 57.
The vertical position.

PAR. 58.
Order litter.

58. Being at the *shoulder:*

1. *Order*, 2. LITTER.

At *order*, the litter is brought to the vertical position; at *litter*, the lower handles are brought to the ground, outside the right foot, canvas to the left, the right arm naturally extended and the hand grasping the front pole; the left hand is dropped to the side.

59. Being at the *order :*

1. *Shoulder,* 2. LITTER.

At *shoulder,* the litter is raised to the vertical position; at *litter,* it is raised vertically until the left wrist is level with the chin, when it is laid, canvas down, upon the shoulder (par. 56), where it is supported by the right arm, the right hand grasping the left pole; the left hand is dropped to the side.

60. Being in line, the litters at the *shoulder,* or *order :*

1. *Carry,* 2. LITTER.

At *carry,* each No. 3 brings his litter to the vertical position; at *litter,* he drops the upper handles forward and downward until the litter is in a horizontal position, canvas to the left; meanwhile the other numbers step directly to the front, No. 2 until he is opposite the front handles, which he seizes with his left hand, and Nos. 1 and 4 until they are opposite the center of the litter. Nos. 2 and 3 take hold by passing the left and right hands respectively outside the handles and grasping the lower one, the litter resting against the hip. The guides step forward and place themselves in line with the front bearers.

PAR. 59.
Shoulder litter.

61. Being at the *carry :*

1. *Ground,* 2. LITTER.

At *ground,* Nos. 2 and 3 face inward, grasping the handles with both hands; at *litter,* they stoop and lower litter to the ground, lengthwise between the files, canvas up, then, standing erect, they face to the front.

62. Being at the *ground :*

1. *Carry,* 2. LITTER.

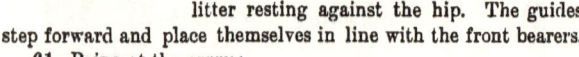

TRANSPORTATION OF THE WOUNDED. 245

PAR. 60.—Carry litter.

At *carry*, Nos. 2 and 3 face inward; at *litter*, they stoop, grasp handles with both hands, and raise the litter from the ground to the *carry*.

63. Being at the *carry:*

1. *Shoulder*, 2. LITTER.

At *shoulder*, No. 3 reaches forward with his left hand and grasps the litter near its center; at *litter*, he brings it to the vertical position and then to the *shoulder;* meanwhile the other numbers step backward and align themselves upon him in regular order.

64. Being at the *carry*, litter *strapped:*

1. *Open*, 2. LITTER.

At *open*, Nos. 2 and 3 face litter; at *litter*, they unbutton the straps and button them to the studs on their respective poles, when they grasp the right (upper) handles, leaving the litter suspended longitudinally, canvas to the left. They then extend the braces and, supporting the litter horizontally by the handles,

PARS. 64 and 68.—At litter posts with open litter.

canvas up, lower it to the ground, and resume the attention, standing between the handles, facing to the front. If the litter be merely closed, they at once grasp the upper handles and proceed as above.

65. Being at the *open*:

1. *Close*, 2. LITTER.

At *close*, Nos. 2 and 3 step respectively outside the right front and left rear handles and face inward; at *litter*, they stoop and raise the litter by the right handles; they then fold the braces and, bringing the lower pole against the upper, canvas to the left, support the litter at the *carry*.

TRANSPORTATION OF THE WOUNDED. 247

66. The litter being *closed:*

1. *Strap,* 2. LITTER.

At *strap,* Nos. 2 and 3 face litter; at *litter,* they fold canvas by doubling it smoothly on top of poles, then pass straps around litter, over canvas, and button them to studs, when they take their posts at the *carry.*

In opening or strapping litter Nos. 1 and 4 may assist when so directed.

PAR. 66.—Strap litter.

In the field, the litter should habitually be carried *strapped* or *closed,* and only *opened* on reaching the patient.

The litter may in like manner be *closed* and then *strapped,* being at the *open,* at the command *strap litter,* when the motions begin with those described under *close litter* (par. 65).

67. To bring the squad into line, the litter being at the *ground* or the *open:*

1. *Form,* 2. RANK.

At *rank*, No. 2 advances one pace, and all align themselves upon him in regular order.

This movement permits the marching of the squad, without litter, to any desired point.

68. Posts at the litter may at any time be recovered by the commands:

1. *At litter*, 2. Posts.

If at the *ground*, the numbers take posts, No. 2 on the right of the front handles, No. 3 on the left of the rear handles and close to them, and Nos. 1 and 4, respectively, on the right and left of the litter at its mid-length and one pace from it; all facing to the front.

If at the *open*, Nos. 2 and 3 take posts between the front and rear handles, respectively, facing to the front, and Nos. 1 and 4 as with litter at the *ground*, but one short step (15 inches) from it.

This is the invariable position taken by each number at the above commands, whatever may have been his previous position or duty.

69. Being *at litter posts*, to change posts:

1. *Change posts*, 2. March.

No. 1 takes No. 3's post, and No. 4 No. 2's, while Nos. 3 and 2 step to the left and right of the litter, respectively, into the vacated positions, all thus describing part of a circle in the same direction around the litter.

70. Being at the *carry* in marching:

1. *Change bearers*, 2. March.

Nos. 1 and 4 step to the right rear and left front of the litter, respectively, and grasp the handles relinquished by Nos. 3 and 2, who step to left and right center respectively.

71. The squad leader continues to exercise command from whatever position he may occupy.

72. To resume the original positions the movement is reversed by the commands:

1. *At litter*, 2. Posts.

73. Being at the *open:*

1. *Prepare to lift*, 2. Lift.

At the first command Nos. 2 and 3 draw loops from the belts, take one in each hand, stoop, slip them upon the handles, and grasp handles. At the second command they slowly rise; No. 4 advances to side of No. 2, and No. 1 steps backward to side of No. 3, adjust slings and observe that everything is right, when they resume their posts.

74. At the commands:

*1. *Forward*, 2. MARCH,

the bearers step off, No. 2 with the left, No. 3 with the right foot, taking short sliding steps of about 20 inches, to avoid jolting and to secure a uniform motion to the litter. Nos. 1 and 4 step off with the left foot.

75. Being at the *lift:*

1. *Lower*, 2. LITTER.

At *litter*, Nos. 2 and 3 slowly lower the litter to the ground, slip loops from the handles, stand erect and pass the loops under their belts.

76. When the litter is to be moved but a few paces, it may be lifted and marched without slings by prefixing *without slings* to the commands: *Prepare to lift, lift.*

77. To carry the litter by four bearers, being at the *open*, the commands are:

1. *By four*, 2. *Prepare to lift*, 3. LIFT.

At the second command Nos. 2 and 3 take posts outside the right front and left rear handles, respectively, and Nos. 4 and 1 outside the opposite handles; they all stoop and grasp handles with both hands. At *lift*, they slowly rise.

78. The *open* litter should be lifted and lowered slowly and without jerk, both ends simultaneously, the rear bearer moving in accord with the front bearer, so as to maintain the canvas horizontal; in fact, the open litter should be handled for purposes of drill as if it were a loaded litter, and as soon as the men are familiar with its manual the drill should, whenever practicable, be with loaded litter.

* The so-called single step, which is by far the easiest for the patient, but which is acquired with difficulty, may also be practiced; No. 2 steps off with the left foot, and No. 3 follows with his right an instant later, and before No. 2 has planted his right; No. 2's right foot next touches the ground, and is immediately followed by No. 3's left.

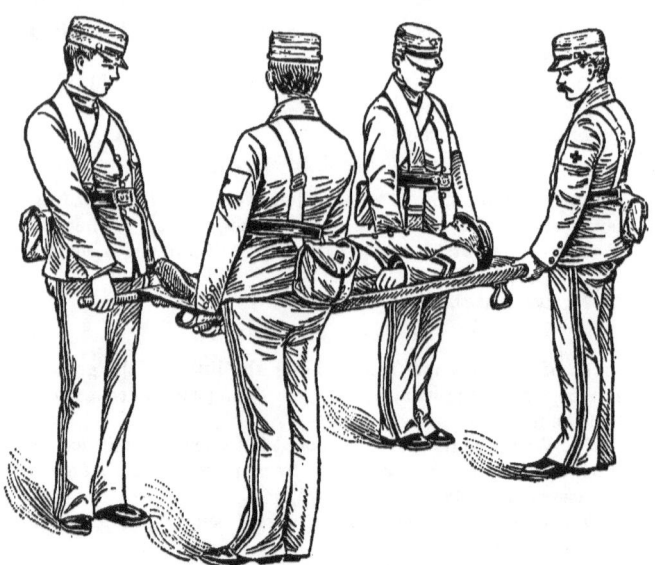

PAR. 77.—Litter lifted by four.

79. Being in line at the *shoulder:*

1. *Return litter*, 2. *Right* (or *left*) *face*, 3. MARCH.

At the first command, the Nos. 3 bring the litter to the vertical position and step one pace to the front; at the second they face as required and bring the litter to the shoulder, and at the third proceed in column of files, by the nearest route, to the place designated for the litters, where they leave them, and, returning in the reverse order, resume their positions by passing through their intervals one pace to the rear and facing about into line. If the place designated is in front of the detachment (or if there be but one squad), the Nos. 3 proceed as described in par. 56. This movement should be supervised by a non-commissioned officer. It can be executed in double time.

MARCHINGS WITH LITTER.

80. The interval between litters, that is, between No. 1 of a squad and No. 4 of the next squad on the right, is 6 inches. It is ordinarily sufficient for the execution of all movements, and should be carefully maintained.

To align a line of litters, at a halt, the litters being at the *carry* or *lift*, the commands are:

1. *Right* (or *left*), 2. DRESS, 3. FRONT.

At *dress*, all dress to the right, the Nos. 2 aligning themselves on the right guide, or No. 2 of the right squad, all promptly recovering their intervals, if lost. At *front*, all face to the front.

81. Being in column, to extend distances, the commands are given:

1. *To two* (or more) *paces, extend*, 2. MARCH.

The first squad advances forward in quick time and the other squads take the short step and successively gain the increased distance; if in march, the first squad, maintains the quick time, while the other squads take the short step as above.

82. The column is closed by the commands:

1. *Litters*, 2. *Close*, 3. MARCH;

when the first squad stands fast (if at a halt), or takes the short step (if in a march), and the other squads successively close up.

83. The line, or column of litters, is marched by the commands already given (par. 24 and following), substituting "*litters*" for "*fours.*"

The following movements require special notice or description:

BEING IN LINE, TO TURN AND HALT.

84. 1. *Detachment*, 2. *Right* (or *left*), 3. MARCH, 4. FRONT.

The first litter halts, and taking the short step, wheels to the right on its own ground; the other litters half wheel to the right and place themselves successively upon the alignment established by the right litter (par. 21).

BEING IN LINE, TO TURN AND ADVANCE.

85. 1. *Detachment*, 2. *Right* (or *left*) *turn*, 3. MARCH.

The first litter takes the short step and wheels to the right on a movable pivot, followed by the others as in par. 22.

BEING IN LINE OF LITTERS, TO MARCH BY THE FLANK IN COLUMN OF LITTERS.

86. 1. *Litters*, 2. *Right* (or *left*), 3. MARCH.

At the command *march*, No. 2 steps off to the right and No. 3 to the left, both describing a quarter of a circle, so as to make the litter revolve horizontally on its center until both face to the right, when they take the full step in the new direction; Nos. 1 and 4 maintain their relative positions opposite the center of the litter. The right guide places himself one pace in front of the first litter, and the left guide one pace in rear of the last litter.

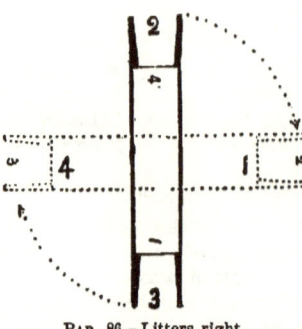

PAR. 86.—Litters right.

BEING IN LINE, TO MARCH IN COLUMN OF LITTERS TO THE FRONT.

87. 1. *Right* (or *left*) *forward*, 2. *Litters right* (or *left*), 3. MARCH.

At the command *march*, the right litter moves straight to the front; the other litters wheel to the right and then to the left in column (par. 29).

To form line from column—see par. 33 and following.

BEING IN LINE OR COLUMN, TO MARCH TO THE REAR.

88. 1. *Litters about*, 2. MARCH.

Nos. 2 and 3 step off as in par. 86, but continue the movement until both face to the rear, the other numbers maintaining their relative positions opposite the center of the litter.

The about with the litter is always to the right.

MOVEMENTS BY SECTIONS.

89. A *section* consists of two litter squads in line.

90. *To form column of sections from line:*

 1. *Sections right* (or *left*), 2. March, 3. Front;

when each section turns as in par. 84; or,

 91. 1. *Sections*, 2. *Right* (or *left*) *turn*, 3. March, 4. *Forward*, 5. March, 6. *Guide right* (or *left*);

when each section turns as in par. 85.

 92. The advantage of this formation is that it permits the shortening of the column, at the *carry*, without increasing its front, by the commands:

 1. *Sections*, 2. *Close*, 3. March;

when the sections close up to one pace, and the litters oblique toward each other until close against the inner free bearers (1 and 4), the outer free bearers meanwhile closing in on their respective sides. In this manner the front is reduced to that of a column of litters.

 93. The normal formation is resumed by the commands:

 1. *Sections*, 2. *Extend*, 3. March.

Line is re-formed by the same commands used to form column.

 94. *To form single column from column of sections:*

 1. *Right* (or *left*) *by litter*, 2. March;

when the second litter of each section takes the short step and obliques to the right behind the first.

 95. *To form column of sections from single column:*

 1. *Form sections*, 2. *Left* (or *right*) *oblique*, 3. March;

when the first squad of each section takes the short step, until the second squad, obliquing to the left, is in line with it.

ROUTE STEP.

 96. The column of strapped litters at the *carry* (par. 60) is the habitual column of route. The rate is 3 to 3½ miles per hour.

Marching in quick time:

 1. *Route step*, 2. March.

The men are not required to preserve silence nor keep the step. The litter squads preserve their distance.

If from a halt:

 1. *Forward*, 2. *Route step*, 3. March.

To resume the cadence step:

1. *Detachment*, 2. ATTENTION.

At the command *attention*, the cadence step in quick time is resumed.

Upon halting while marching in route step, the men come to the rest at the *ground* (par. 61).

97. To march at ease:

1. *At ease*, 2. MARCH.

The detachment marches as in the route step, except that silence is preserved.

THE LOADED LITTER.

TO LOAD AND UNLOAD THE LITTER.

98. For drill in loading litter, the "patients" are directed to lie down, at suitable intervals, near the line of litters, first with head and later with feet toward it, and lastly in any position. Each squad may be separately exercised under its leader or an instructor, or several squads simultaneously.

99. The litters being at the *open*, the instructor commands:

1. *At patient*, 2. *Right* (or *left*), 3. POSTS.

If the command is *right*, Nos. 2, 1, and 3 take positions, No. 2 at the right ankle, No. 1 at the right hip, and No. 3 at the right shoulder, while No. 4 takes position by the left hip opposite No. 1, all facing the patient.

If the command is *left*, Nos. 2, 4, and 3 take position, No. 2 at the left ankle, No. 4 at the left hip, and No. 3 at the left shoulder, while No. 1 takes position at the right hip. opposite No. 4, all facing the patient.

It will be seen from the above that, whether the command is *right* or *left*, the positions of Nos. 1 and 4 are invariable, No. 1 at the right hip, No. 4 at the left hip, and that the positions of Nos. 2 and 3 are always at the ankle and shoulder, respectively, on the right or left of the patient, as the command may be; if *right*, they are on each side of No. 1; if *left*, they are on each side of No. 4.

These movements assume that the patient is lying on his back; but as in the field he will often lie on his face or side, the

TRANSPORTATION OF THE WOUNDED.

bearers should be practiced in promptly taking their proper positions under all circumstances.

These positions are taken by the bearers, whatever may have been their previous positions or duties.

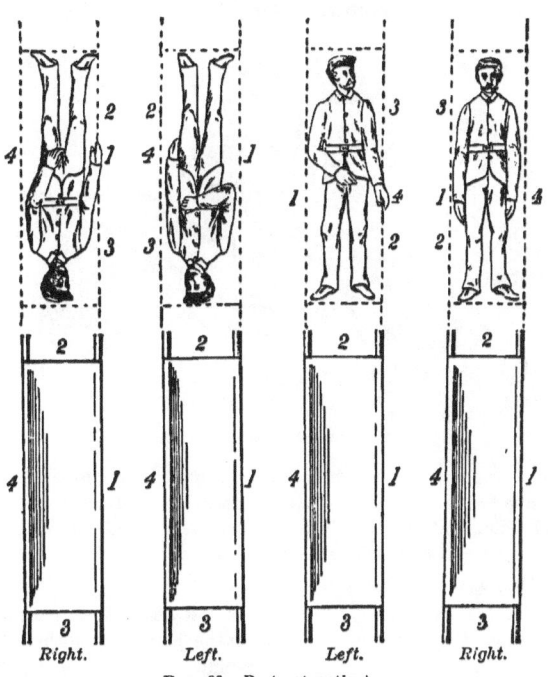

PAR. 99.—Posts at patient.

100. 1. *Prepare to lift*, 2. LIFT.

At the first command all the bearers kneel on the knee nearest the patient's feet (right knee if on the right of the patient, and on the left knee if on his left); No. 2 passes both forearms under the patient's legs, carefully supporting the fracture, if there be one; Nos. 1 and 4 pass their arms under the small of his back and thighs, not locking hands; No. 3 passes one hand under his neck to the further armpit, with the other supporting the nearer shoulder.

At the second command all lift together slowly and carefully and place the patient upon the knees of the three bearers. As soon as he is firmly supported there, the bearer on the free side

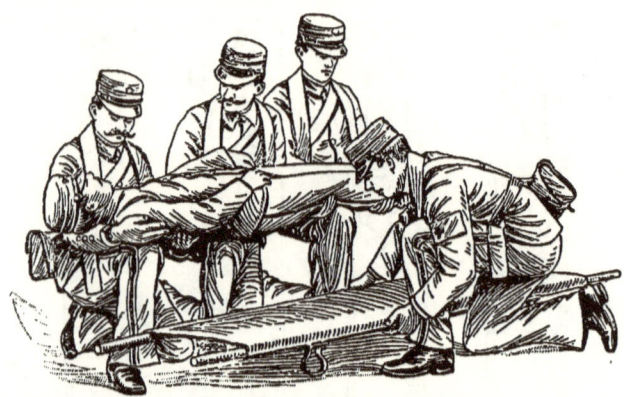

PAR. 100.—The patient lifted.

(No. 1 or 4) relinquishes his hold, passes quickly and by the shortest line to the litter, which he takes up by the middle, one pole in each hand, and returning rapidly places it under the patient and against the bearers' ankles.

101. 1. *Lower*, 2. PATIENT.

The free bearer, No. 1 or 4, stoops and assists the other numbers in gently and carefully lowering the patient upon the litter. The bearers then rise and at once resume their positions *at litter posts* (par. 68).

102. To unload, posts are taken at patient as in loading. At the commands:

1. *Prepare to lift*, 2. LIFT (Par. 100),

they raise him upon the knees, the free bearer removes the litter, and at *lower patient*, they lower him carefully to the ground.

103. In the field, Nos. 1 and 4 going ahead, having reached the patient and taken position on their respective sides, remove his arms and accouterments and examine him to determine the site and nature of the injury; they administer restoratives, if required, and apply such dressings or splints as are needful or available. As soon as Nos. 2 and 3 reach the patient, they help, as far as may be necessary, in rendering this first aid.

104. The drill should be made as nearly as possible like service in actual warfare. For this purpose a diagnosis tag having been attached to the clothing of the "wounded," indicating the site and character of the injury to be dressed before loading, they are directed to take positions at variable distances, in or out of sight, such as they would occupy on the battlefield.

105. The litter being at the *carry*, at the commands:

   1. *Search for wounded*, 2. MARCH;

each leader assumes charge of his squad and proceeds independently. Nos. 1 and 4 at once start ahead to search, but without losing sight of the litter, which follows in quick time, taking the double time as soon as a patient is discovered. The litter is halted and opened (by No. 3's commands) in the most convenient position near the patient. The injury having been dressed, No. 1 commands:

106. 1. *At patient*, 2. *Right* (or *left*), 3. POSTS.

As a rule, the command should be right or left, according as the right or left side of the patient is injured, so that by having the three bearers on that side a better support may be given to the wounded parts.

107. In the field, when the ground on which the patient lies is such that the litter can not be placed directly under him, it should be placed as near him as possible, and preferably in a direction parallel to, or in line with him, when it will be necessary to carry the patient to the litter, instead of the litter to the patient. In such case, the bearers having brought the patient upon their knees, as described in par. 100, at the command *rise*, take firm hold of him and rise, and at *march*, carry him forward, or by the flank as directed. From this position he is first lowered to the knees of the bearers, and thence placed upon the litter or ground.

108. At the commands:

   1. *Cease*, 2. DRILLING,

the squads re-form in line and lower litters, when the patients, if still upon the litters (the dressings, if any, having been removed), are directed to rise and resume their posts, after which the litters are strapped.

POSITION OF PATIENT ON THE LITTER.

109. The position of a patient on the litter depends on the character of his injury. An overcoat, blanket, knapsack, or other suitable and convenient article, should be used as a pillow to give support and a slightly raised position to the head. If the patient is faint, the head should be kept low. Difficulty of breathing in wounds of the chest is relieved by sufficient padding underneath. In wounds of the abdomen the best position is on the injured side, or on the back if the front of the abdomen is wounded, the legs in either case being drawn up, and a pillow or other available object placed under the knees to keep them bent.

In an injury of the upper extremity, calling for litter transportation, the best position is on the back, with the injured arm laid over the body or suitably placed by its side, or on the uninjured side, with the wounded arm laid over the body. In injuries of the lower extremity the patient should be on his back, or inclining toward the wounded side; in cases of fracture of either lower extremity, if a splint can not be applied, it is always well to bind both limbs together.

GENERAL DIRECTIONS.

110. In moving the patient either with or without litter, every movement should be made deliberately and as gently as possible, having special care not to jar the injured part. The command *steady* will be used to prevent undue haste or other irregular movements.

111. *The loaded litter should never be lifted or lowered without orders.*

112. The rear bearer should watch the movements of the front bearer and time his own by them, so as to insure ease and steadiness of action.

113. The number of steps per minute will depend on the weight carried and other conditions affecting each individual case.

TRANSPORTATION OF THE WOUNDED. 259

114. The handles of the litter should be held in the hands at arm's length and supported by the slings. Only under the most exceptional conditions should the handles be supported on the shoulders.

115. The bearers should keep the litter level notwithstanding any unevenness of the ground.

116. In making ascents or descents, the rear or front handles should be raised to bring the litter to the proper level, Nos. 1 and 4 supporting the poles on their respective sides. With only three bearers, the free bearer takes hold of the handle on his side.

117. As a rule, the patient should be carried on the litter feet foremost, but in going up hill his head should be in front; in case of fracture of the lower extremities, he is carried up hill feet foremost, and down hill head foremost, to prevent the weight of the body from pressing down on the injured part.

TO PASS OBSTACLES.

118. A breach should be made in a fence or wall for the passage of the litter, if there be no gate or other opening, but

PAR. 118.—Passing an obstacle.

should it be necessary to surmount the obstacle, the commands are given: **1.** *Obstacle*, **2.** MARCH.

At *obstacle*, Nos. 1 and 4 close in to side of litter, grasp poles with both hands and support it; at *march*, No. 2 slips loops from handles and, climbing over, receives litter as it advances (facing to the front); Nos. 1 and 4 then pass the obstacle and resume their places at the poles when the litter is carried over; No. 3, slipping loops from handles, now also climbs over and takes his place between the handles; the slings having been adjusted without halting, Nos. 1 and 4 resume their posts.

119. The passage of a narrow cut or ditch is effected in a similar manner; Nos. 1 and 4 bestride or descend into the cut, support and advance the litter until No. 2 has crossed and resumed his hold, when the litter is carried over; No. 3 then crosses and all resume their places.

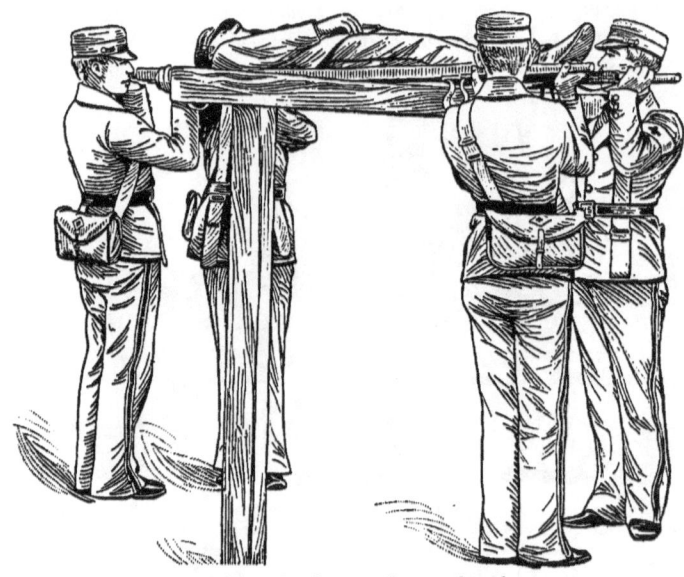

PAR. 121.—Surmounting an obstacle.

TRANSPORTATION OF THE WOUNDED. 261

120. In crossing a running stream, or broken or otherwise difficult ground, or going up or down hill, Nos. 1 and 4 give support on their respective sides of the litter at the command *obstacle* (par. 118); or the litter may be carried *by four* (par. 77).

121. To surmount an obstacle over 5 feet high, the litter being lifted *by four* (par. 77), the commands are given:

1. *Raise*, 2. LITTER;

when the litter is carefully raised to the level of the obstacle and passed over until the front legs have cleared it, where it is rested; Nos. 2 and 4 cross the obstacle and resume hold of their handles on the other side; the litter is then advanced until only the rear handles rest upon the obstacle, when Nos. 1 and 3 get over and resume hold of their handles; the litter is then lowered.

122. To pass a wide cut or ditch, the litter being lowered with the front legs near the edge, Nos. 2 and 4 descend into the cut, take hold of the front handles, and the movement proceeds as in par. 121.

TO CARRY A LOADED LITTER UPSTAIRS.

. 123. A loaded litter is usually carried upstairs head front, and downstairs feet front, as follows: The litter is marched to the foot of the stairs in the usual manner, wheeled about, halted, and lowered, when the commands are given:

1. *Prepare for stairs*, 2. LIFT, 3. MARCH.

At the first command, No. 4 steps to the handle nearest him at the foot of the litter, which he grasps firmly with both hands, while No. 2 grasps the opposite handle; No. 3 faces about, stoops, replaces loops, and grasps handles. At *lift*, the litter is lifted, and at *march*, carried up by Nos. 3, 2, and 4, the rear bearers (Nos. 2 and 4) keeping it as level as possible by raising their end, and, if necessary, resting the handles on shoulders. They must carefully watch the patient to prevent accident. In this movement No. 1 falls out and accompanies litter, to render any assistance required.

If the litter is to be carried any considerable distance, it must be lowered as soon as possible after the stairs are mounted, and the usual positions resumed.

PAR. 123.—Carrying a loaded litter upstairs.

TO CARRY A LOADED LITTER DOWNSTAIRS.

124. As the litter approaches the stairs, the command, *Prepare for stairs*, is given, when No. 4 advances to left front handle, which he grasps firmly with both hands, while No. 2 grasps the opposite handle. The front bearers, as the descent proceeds, bring handles to shoulders, if necessary, to keep the litter as level as possible; the rear bearer carefully observes patient to see that no accident happens to him. When the level is reached the usual positions are resumed. When necessary, the litter may be lowered at the head of the stairs.

When, for any reason, it is necessary to carry a patient feet first upstairs or head first downstairs, the bearers are simply reversed, in the former case No. 2 becoming front bearer, and in the latter No. 3, who is assisted by No. 1.

TRANSPORTATION OF THE WOUNDED.

PAR. 124.—Carrying a loaded litter downstairs.

TO LOAD AND UNLOAD WITH REDUCED NUMBERS.

125. In loading with reduced numbers, No. 2 or 3 (absent) is ordinarily replaced by No. 4 or 1, respectively, while Nos. 1 and 4 replace each other (par. 52).

With three bearers the litter is placed as usual, and at the prescribed commands the bearers take their proper positions. The patient having been lifted by the three bearers, is supported on the knees of the two on one side, while the third (No. 1 or 4) places the litter in position under him.

126. Another method for three bearers, when it is necessary to carry the patient to the litter, is as follows: Two bearers take their positions on the same side opposite the knee and hip, while the third stands by the opposite hip. At the usual commands the two bearers at the hips stoop, and raising the patient to a sitting position, place one hand and arm around the back and interlock the fingers of the other hand, palms up, under the upper

PAR. 126.—Carrying by three bearers.

part of the thighs. The patient, if able, clasps his arms around their necks. The third bearer (No. 2 or 4) supports the lower extremities with both arms passed under them, one above, the other below the knee.

127. If only two bearers are available (*e. g.*, Nos. 2 and 3), the patient is necessarily always carried to the litter. No. 2 proceeds by the right and No. 3 by the left, and take position on opposite sides of the patient near his hips. They lift the patient as directed (par. 145), the legs remaining unsupported, and carry him feet foremost over the near end of the litter.

In case of fractured lower extremity, the two bearers kneel on the injured side, raise the patient upon their knees, and take firm hold of him, No. 2 supporting both lower extremities, while No. 3 supports the body, the patient clasping his arms around his neck, when they rise.

128. To unload with three bearers, posts are taken at patient as in loading. At the commands:

1. *Prepare to lift*, 2. Lift (Par. 100),

they raise him upon the knees, No. 1 (or 4) removes the litter, and at *lower patient*, they lower him carefully to the ground.

129. With two bearers, they form a two-handed seat (par. 145), and lift the patient off the litter. In the case of fracture, they stand on the same side, and kneeling (par. 127), lift him upon their knees, then rise and take two steps backward to clear the litter, when they lower him to their knees and then to the ground.

TO TRANSFER PATIENT FROM LITTER TO BED OR ANOTHER LITTER.

130. *From litter to bed:* With four or three bearers the litter is placed at the foot of the bed, as nearly as possible in line with it; the bearers taking their positions (all on one side if only three), lift the patient upon their knees, then, at the command *rise*, taking firm hold of him, they rise, and, moving cautiously by side steps to the bedside, lower him upon the bed.

With two bearers, the patient is likewise first lifted upon their knees, then carried by side steps to the bed.

131. *From litter to litter:* The patient is lifted upon the knees of the bearers, the litter removed and replaced by the other litter.

IMPROVISATION OF LITTERS.

132. Many things can be used for this purpose:

Camp cots, window shutters, doors, benches, boards, ladders, etc., properly padded.

Litters may be made with sacks or bags of any description, if large and strong enough, by ripping the bottoms and passing two poles through them and tying crosspieces to the poles to keep them apart; two, or even three, sacks placed end to end on the same poles may be necessary to make a safe and comfortable litter.

Bedticks are used in the same way by slipping the poles through holes made by snipping off the four corners.

Pieces of matting, rug, or carpet trimmed into shape, may be fastened to poles by tacks or twine.

Straw mats, leafy twigs, weeds, hay, straw, etc., covered or not with a blanket, will make a good bottom over a framework of poles and cross sticks.

Better still is a litter with bottom of ropes or rawhide strips whose turns cross each other at close intervals.

133. But the usual military improvisation is by means of rifles and blankets. Each squad should be supplied with two rifles (bayonets fixed), carried by Nos. 1 and 3, who assure themselves

PAR. 133.—Preparing blanket litter.

that they are unloaded, and a regulation blanket rolled up, and carried by No. 4 over right shoulder. The detachment being in line, is formed in column of fours (par. 25), when the commands are given:

1. *For blanket litter*, 2. MARCH.

At *march*, Nos. 2 and 4 step two paces forward, and Nos. 1 and 3 place themselves behind them.

1. *Prepare*, 2. BLANKET LITTER.

Nos. 2 and 4 face about; No. 4 takes blanket roll and passes one end to No. 2, when Nos. 1 and 3 seize free edge of blanket (with free hands) as near the corners as possible; Nos. 2 and 4 step backward till the blanket is unrolled, when all stoop and place it smoothly on the ground. Nos. 1 and 3 lay rifles under edges of blanket, trigger guard in, muzzles toward 2 and 4, somewhat converging, when all roll blanket tightly about rifles, an equal number of turns on each piece, until the space between them measures 20 inches, and stand erect.

134. The blanket litter is lifted *by four* as in par. 77. It may be carried in any direction, and all movements of loading, unloading, etc., are executed as laid down for the hand litter.

135. When no longer required, the commands are given:

1. *Take apart*, 2. BLANKET LITTER.

The bearers having resumed their original positions, face the litter, stoop and unroll blanket on their respective sides; Nos. 1 and 3 take up the rifles and stand at the *order;* Nos. 2 and 4 fold the blanket lengthwise, then roll it tightly, when No. 4 places the roll upon his shoulder and all stand facing to the front.

At *form, rank*, each squad is re-formed as in par. 67. The column of fours may now be re-formed into line (par. 33).

136. Should it be desirable, by reason of the patient's condition, or because of reduced numbers of the squad, the following method may be used:

One half of the blanket is rolled lengthwise into a cylinder, which is placed along the back of the patient, who has been turned carefully on his side. The patient is then turned over upon the blanket and the cylinder unrolled on the other side. The rifles are then laid down and rolled tightly in the blanket, each a like number of turns, until the side of the body of the patient is reached, when they are turned trigger guards up.

137. A litter may also be prepared with two rifles and two or three blouses, by turning the blouses lining out, and buttoning them up, sleeves in, when the rifles are passed through the sleeves, the backs of the blouses forming the bed.

Two bearers may carry the wounded man in these improvisations, but it is better, whenever possible, that four men should do so, two on each side.

METHODS OF REMOVING WOUNDED WITHOUT LITTERS.

FOR ONE BEARER.

138. While it is not desirable that one bearer should, ordinarily, be required or permitted to lift a patient unassisted, emergencies may arise when a knowledge of proper methods of lifting and carrying by one bearer is of the utmost value.

A single bearer may carry a patient in his arms, on his back, or across his shoulder.

To bring the patient into any of these positions, the first steps are as follows:

139. *To lift the patient erect.*

The bearer, turning patient on his face, steps astride body, facing toward the head, and with hands in his armpits lifts him to his knees, then clasping hands over the abdomen, lifts him to his feet; he then with the left hand seizes the patient by the left wrist and drawing the left arm about his (the bearer's)

Par. 139.—Lifting patient erect.

TRANSPORTATION OF THE WOUNDED.

neck holds it against his left chest, the patient's left side resting against his body, and supports him with his right arm about the waist.

140. From this position the bearer proceeds as follows:

To lift the patient in arms.

The bearer, with his right arm behind patient's back, passes his left under thighs and lifts him into position.

To place patient astride of back.

141. The bearer shifts himself to the front of patient, back to him, stoops, and, grasping his thighs, brings him well up on his back.

As the patient must help himself by placing his arms around the bearer's neck, this method is impracticable with an unconscious man.

To place the patient across back.

142. The bearer with his left hand seizes the right wrist of the patient and draws the arm over his head and down upon his left shoulder, then shifting himself in front, stoops and clasps the right thigh with his right arm passed between the legs, his right hand seizing the patient's right wrist; lastly, he, with his

PAR. 142.—Patient across back.

left hand, grasps the patient's left and steadies it against his side, when he rises.

To place the patient across shoulder.

143. The bearer clasps his hands about the patient's waist, shifts himself to the front, facing him, and stooping places his right shoulder against the abdomen; he passes his right hand and

PAR. 143.—Patient across shoulder.

arm between the thighs—securing the right thigh—and with his left grasps patient's right hand, bringing it from behind under his (bearer's) left armpit, when, the wrist being firmly grasped by his right hand, he rises. This position leaves the left hand free.

TRANSPORTATION OF THE WOUNDED. 271

144. In lowering patient from these positions the motions are reversed. Should a patient be wounded in such manner as to require these motions to be conducted from his right side, instead of left, as laid down, the change is simply one of hands—the motions proceed as directed, substituting right for left, and *vice versa*.

FOR TWO BEARERS.

By the two-handed seat.

145. The patient lying on the ground, the commands are given:
1. *Form two-handed seat*,
2. *Prepare to lift*,
3. LIFT.

At the first command the two bearers take positions facing each other on the right and left of the patient near his hips.

At the second command they raise the patient to a sitting posture, pass each one hand and arm around his back, while the other hands are passed under the thighs, palms up, and the fingers interlocked.

At *lift*, both rise together.

In marching, the bearers should break step, the right bearer starting with the right foot, the left bearer with the left foot.

By the extremities.

146. This method requires no effort on the part of the patient; but it is not applicable to

PAR. 145.—Two-handed seat.

severe injuries of the lower extremities. One bearer stands by the patient's head, the other between his legs, both facing toward the feet. At *prepare to lift*, the rear bearer having raised the patient to a sitting posture, clasps him from behind around the body under the arms, while the front bearer, standing between the legs, passes his hands from the outside under the flexed knees. At *lift*, both bearers rise together.

By the rifle seat.

147. A good seat may be made by running the barrels of two rifles through the sleeves of an overcoat, buttoned as in par. 187, so that the coat lies back up, collar to the rear. The front bearer rolls the tail tightly around the barrels and takes his grasp over them; the rear bearer holds by the butts, trigger guards up.

148. A stronger seat is secured in the following manner:

A blanket being folded once from side to side, a rifle is laid upon it transversely across its center so that the butt and muzzle project beyond edges; one end of the blanket is folded upon the other end and a second rifle laid upon the new center in the same manner as before. The free end of the blanket is folded upon the end containing the first rifle so as to project a couple of inches beyond the first rifle. The litter is raised from the ground with trigger guards up.

TO PLACE A PATIENT ON HORSEBACK.

149. The help required to mount a disabled man will depend on the site and nature of his injuries; in many cases he is able to help himself materially. If he be entirely helpless, five men—if available—should be used to mount him, one to hold the horse, the others to act as bearers. The horse is, if necessary, blindfolded.

To load from the near side, the bearers take posts *at patient left*, lift the patient, and at the command *prepare to mount* carry him to horse, his body parallel to that of the horse, his head toward the horse's tail. No. 1 stands on the off side of the horse ready to grasp the right leg of the patient when it is brought within his reach. When close to the horse's side, at the command *mount*, the patient is carefully raised and carried over the horse until his seat reaches the saddle, when he is raised into position.

Par. 149.—Placing patient on horseback.

When necessary to load from the off side the bearers take post *at patient right*.

To mount with the assistance of three or two bearers the same principle is observed.

150. To dismount, at the commands,

1. *At patient*, 2. *Right* (or *left*), 3. Posts, 4. *Prepare to dismount*, 5. Dismount,

the movements are reversed.

151. The patient once mounted should be made as safe and comfortable as possible. A comrade may be mounted behind him to hold him and guide the horse; otherwise, a lean-back must be provided, made of a blanket roll, a pillow, or a bag filled with leaves or grass. If the patient be very weak, the lean-back can be made of a sapling bent into an arch over the cantle of the saddle, its ends securely fastened, or of some other framework to which the patient is bound.

THE TRAVOIS.

152. The travois is a vehicle intended for transporting the sick or wounded when the use of wheeled vehicles or other means of transportation is impracticable. It consists of a frame, having shafts, two side poles and two crossbars, upon which a litter may be rested and partly suspended. When in use, a horse or mule is attached to the shaft and pulls the vehicle, the poles of which drag on the ground. One pole is slightly shorter than the other, in order that in passing an obstacle the shock may be received successively by each and the motion be equably distributed.

153. No. 4 procures the animal, sees that it is properly harnessed, and keeps charge of it. The travois having been procured by Nos. 2 and 3, the commands are given:

154. 1. *Prepare*, 2. Travois.

Nos. 1 and 3 take posts (left and right respectively) at the ring (or front) ends of the shafts, and No. 2 at the shoe (or rear) ends of the side poles. They pull the shafts forward through the collars until fully extended; then Nos. 1 and 3 place the front crossbar over the front end of the side poles, driving it home until its collars strike the front collars of the side poles, and No. 2 passes the collars of the rear crossbar (keeping uppermost the surface on which are the flat bolts) over the rear ends of the poles, pushing them forward until they reach the squared places, beyond the bolt slots, when the barrel bolts are thrown into place.

155. 1. *Hitch*, 2. Travois.

Nos. 1 and 3 hitch the travois. If the animal has an ordinary wagon harness, the rings on the front ends of the shafts are put over the iron hooks on the hames, and the toggle of each trace chain is fastened to the ring of the corresponding side pole. · If it is saddled, the rings on the front ends of the shafts are fastened to the pommel of the saddle by means of the straps that belong there, and the shafts secured by a surcingle passed over all.

The saddle should have a breast strap, if practicable.

156. Nos. 1, 2, and 3 having taken their posts at litter, the travois is loaded as in ambulance drill (pars. 169 and 170), by the commands:

1. *Take post to load travois*, 2. March.

1. *Prepare to load*, 2. Load.

At *load*, the three bearers slowly raise the litter and carry it lengthwise over the travois. Nos. 1 and 3 slip the handles of the litter through the leather loops on the front ends of the side poles, and set the front legs into the mortises, securing them by the bolts, when they take posts at travois, facing the animal, No. 1 at the patient's right shoulder, No. 3 at his left, and No. 2 at his feet.

157. The squad being at travois posts, at the commands:

1. *Prepare to unload*,
2. UNLOAD,

the movements are reversed and the litter lowered clear of the travois.

158. The travois may be loaded or unloaded by two bearers who take posts opposite each other at mid-length of litter which they lift on or off the travois, grasping each pole with both hands.

159. The travois being unloaded, at the commands:

1. *Unhitch*, 2. TRAVOIS,

the travois is unhitched by Nos. 1 and 3, and the animal led out by No. 4.

1. *Pack*, 2. TRAVOIS.

Nos. 1, 2, and 3 at their respective posts remove the rear and front crossbars, then slip the shafts backward under the side poles and secure all by the cross straps.

160. A travois may be improvised by cutting poles about 15 feet long and 2 inches in diameter at the small end. These poles

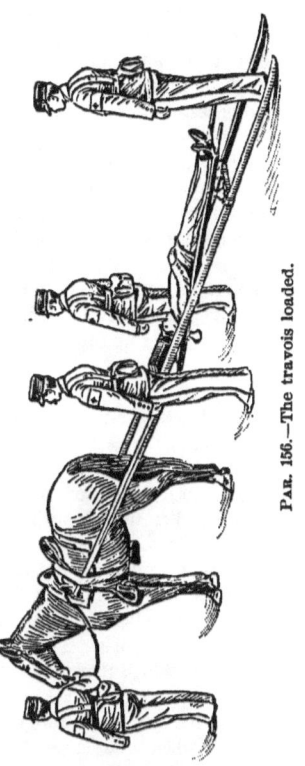

PAR. 156.—The travois loaded.

are laid parallel to each other, small ends to the front and 2½ feet apart; the large ends about 3 feet apart, and one of them projecting 8 or 10 inches beyond the other. The poles are connected by a crossbar about 6 feet from the front ends and another about 6 feet back of the first, each notched at its ends and securely lashed at the notches to the poles. Between the crosspieces the litter bed, 6 feet long, is filled in with canvas, blanket, etc., securely fastened to the poles or crossbars, or with rope, lariat, rawhide strips, etc., stretching obliquely from pole to pole in many turns, crossing each other to form the basis for a light mattress or improvised bed; or a litter may be made fast between the poles to answer the same purpose. The front ends of the poles are then securely fastened to the saddle of the animal. A breast strap and traces should, if possible, be improvised and fitted to the horse. On the march, the bearers should be ready to lift the rear end of the travois, when passing over obstacles, crossing streams, or going up hill.

THE TWO-HORSE LITTER.

161. *The two-horse litter* consists of a litter with long handles used as shafts for carrying by two horses, or mules, one in front, the other in rear of the litter. It accommodates one recumbent patient. On a good trail it is preferable to the travois, as the patient lies in the horizontal position, and in case of fractured limbs they can easily be secured against disturbance. This litter may be improvised in the same manner as the travois, only the poles should be 16½ feet long, and the crossbars forming the ends of the litter bed should be fastened 5 feet from the front and rear ends of the poles. The ends are made fast to the saddles by notches, into which the fastening ropes are securely tied.

162. A patient is placed upon a horse litter after it is hitched, and removed from it before it is unhitched, in the same general manner as when lowered upon or lifted from a bed or other litter.

THE AMBULANCE.

163. The regulation ambulance is a four-wheeled vehicle, drawn by two horses. It provides transportation for eight men sitting or two recumbent on litters, or four sitting and one recumbent. It is fitted with four removable seats, which, when not used as such, are hung, two against each side, thus answering the purpose of cushions. The floor is 7½ feet long and 4 feet wide. Beneath the driver's seat is a box for the medical and surgical chests, and under the body are two water tanks; outside, on each side, are two brackets upon which litters are carried.

AMBULANCE DRILL.*

164. The litters are said to be *packed* when they are strapped and placed upon the brackets. The seats are said to be *prepared* when they are horizontal, supported by the legs; and *packed* when they are hooked against the sides of the wagon.

165. Being in line:

1. *At ambulance*, 2. Posts.

The designated squad marches in column of files to the ambulance; when No. 1 takes post on the left, No. 2 in the center, and

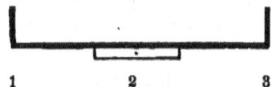

1   2   3   4

No. 3 on the right of the rear of the ambulance and close to it, No. 4 on the right of No. 3.

In the case of a litter lowered in rear of ambulance preparatory

---

* With the old-pattern ambulance, which has a litter rest and special litters (the latter carried suspended each by two straps from the hand rails), and which will not accommodate the regulation litter, the bearers proceed as follows: After the seats are *packed*, No. 2 passes the two rear rollers to No. 3 and adjusts the two front rollers; they then, beginning on the right, unbuckle the straps, place the litter upon the rollers, and resume their posts. No. 1 then commands: *Procure litter, about face*, when the litter is brought to the *carry* and *opened;* after which the patient is transferred to the ambulance litter and *loaded*. After the patient is *unloaded* and transferred to the regulation litter, No. 1 commands: *At ambulance litter posts*, when the litter is *closed, shouldered*, and *returned* to ambulance; he then marches his squad back to the regulation litter. To *prepare seats*, the litters are suspended, beginning on the right, the rollers are returned to their compartment, and the seats *prepared*.

to loading, head of patient toward it, at the command *posts*, each No. faces about and proceeds directly to his post.

This is the invariable position of the squad *at ambulance posts;* it may be taken from any position (the litter, if any, being *grounded* or *lowered*), and when disarranged, from whatever cause, the squad may be reassembled by these commands for service at the ambulance.

166. The ambulance having seats *packed* and the squad being *at ambulance posts:*

1. *Prepare,* 2. SEATS.

Nos. 1 and 3 raise the curtain, if necessary, and open the tail gate; Nos. 2 and 3 enter the ambulance, No. 2 facing the front and No. 3 the rear seat of their respective sides. Each man seizes the lower edge of the seat about 6 inches from the ends with both hands and lifts it carefully to free the hooks from the upper slots, and then slips them into the lower slots; he raises the legs and adjusts them to the seat, fastening the keepers, and tries the seat for firmness before leaving it. He then prepares in like manner the opposite seat. Nos. 3 and 2 now resume their places *at ambulance posts.*

167. The ambulance having seats *prepared*, and the squad being *at ambulance posts:*

1. *Pack,* 2. SEATS.

Nos. 1 and 3 raise the curtain, if necessary, and open the tail gate, Nos. 2 and 3 enter the ambulance, No. 2 facing the front and No. 3 the rear seat of their respective sides. Each man having pushed aside the keepers covering the slots, releases the legs, then seizing the front of the seat with both hands, raises the seat to clear the hooks from the lower slots and slips them into the upper slots. He then packs in like manner the opposite seat, Nos. 3 and 2 now resume their posts at ambulance.

168. Seats can be *prepared* or *packed* on one side only (leaving room on the packed side for one recumbent patient) by the commands: *Prepare* (or *pack*) *seats, right* (or *left*).

169. The litter being *lifted,* at the commands:

1. *Take post to load ambulance,* 2. MARCH;

the squad proceeds to the ambulance. No. 4, starting ahead in double time, lays the arms and accouterments of the patient

TRANSPORTATION OF THE WOUNDED. 279

(which he carries) on the ground by the right rear wheel; he then raises the curtain, if necessary, opens the tail gate, observes the condition of the ambulance and (resuming his post at the litter) reports it to the squad leader. Upon approaching the ambulance the litter is wheeled about so that the head of the patient is toward the rear of the ambulance and two paces from it, when the litter is halted and lowered. If it be necessary to prepare the ambulance before loading, the squad takes posts at ambulance, No. 4 remaining in charge of the patient; if ready for the reception of the litter the commands are given:

1. Prepare to load, 2. LOAD.

170. At the first command No. 2 faces about, No. 3 steps around his left handle and takes post at the patient's left shoulder; No. 1 takes post opposite No. 3; all facing the litter stoop, No. 2 grasping his handles, and Nos. 1 and 3 their respec-

PAR. 170.—Loading ambulance.

tive poles; No. 4 watches the patient and otherwise renders any needful assistance. At *load*, the bearers slowly raise the litter to the level of the floor of the ambulance and advance to it, being careful to keep the litter in a horizontal position; the legs are placed on the floor by Nos. 1 and 3, and the litter pushed in by No. 2 assisted by the others. When this is accomplished Nos. 1, 2, and 3 are in position *at ambulance posts*. No. 4 places the arms and accouterments of the patient (if any) under the litter, and then takes his position on the right; Nos. 1 and 3 close the tail gate and, if necessary, lower the curtain. The squad may then be faced in any desired direction and marched away.

171. The squad being *at ambulance posts:*

1. *Prepare to unload*, 2. UNLOAD.

At the first command Nos. 1 and 3 raise the curtain, if necessary, open the tail gate, and No. 2 takes hold of the projecting handles of the litter; at *unload*, No. 2 draws out the litter, assisted by Nos. 1 and 3, who, facing inward, support the poles until the inner handles are reached. The litter, carefully supported in a horizontal position, is then lowered with the head of the patient two paces in rear of wagon; No. 4 closes the tail gate and all take posts at litter.

172. To load with two bearers, the litter being lowered in position for loading, at *prepare to load*, the bearers take posts on their respective sides, mid-length of the litter and facing it; they stoop and grasp each a pole firmly with both hands. At *load*, they lift the litter and push it into the ambulance.

At *unload*, each bearer grasping his handle, they partly withdraw the litter, then shifting their hands to their respective poles and facing each other they continue to withdraw it until the head reaches the rear of the ambulance, when they lift the litter out and lower it to the ground.

173. The right side of the ambulance is always loaded or unloaded first, unless otherwise ordered.

174. When necessary to load the feet first, at the commands: *By the feet, take post to load ambulance, march*, the litter is lowered with foot toward the ambulance, when the loading proceeds as above described, excepting that No. 3 remains between his handles, No. 2 takes post opposite the right ankle, and No. 4 opposite him.

175. At the conclusion of the drill with ambulances the detachment is re-formed in line.

TO PREPARE AND LOAD ORDINARY WAGONS TO TRANSPORT WOUNDED.

176. In active service, the use of the ordinary army or other wagons for transporting the sick and wounded is of everyday occurrence, and it is important that bearers should be practiced in preparing, loading, and unloading such vehicles. Patients may be laid on straw or other light material spread thickly over the bottom of the wagon, or on hand litters placed on the bottom, or suspended by ropes or straps. The movements heretofore fully described, to *load* and *unload*, will, if thoroughly understood, meet the requirements of any emergency of this character. It must, however, always be remembered that such work demands a far greater amount of care on the part of the bearers, for the safety and comfort of their patients, than when the proper appliances are at hand.

INSPECTION AND MUSTER.

INSPECTION OF DETACHMENT.

177. Inspection is in such uniform as may be prescribed. The Hospital Corps pouch is worn with all uniforms, suspended from the left shoulder to the rear over the right hip.

The detachment should be frequently inspected in the uniform and equipment for field service, which consist of the blouse, trousers, campaign hat, shoes, and leggings, the waist belt, pouch and sling, the haversack and canteen suspended from the right shoulder to the rear over the left hip, and the tin cup hung from the flap strap of the haversack.

If required, a revolver is carried at the belt on the right side and a cartridge pouch on the left.

178. The detachment being formed, the senior non-commissioned officer salutes, reports, and takes his place on the right of the line of file closers (par. 3).

The officer commanding, standing in front of the center of the detachment, then draws sword and commands:

    1. *Open ranks*, 2. MARCH, 3. FRONT.

At the first command the senior non-commissioned officer steps one pace to the rear to mark the new alignment of the file closers; the medical officers above the rank of lieutenant stand fast during the inspection; the junior officers place themselves on the right and left of the rank; the officer commanding goes to the right flank and verifies the position of the senior non-commissioned officer, then places himself facing to the left, three paces in front of the right of the detachment, and commands: MARCH. At this command the junior officers take post three paces in front of the detachment, distributing themselves equally along the line, in order of rank, from right to left; the rank (the left hand above the hip) dresses to the right; the file closers step backward to the line established by the senior non-commissioned officer, and dress to the right.

The officer commanding aligns the officers and the rank; the senior non-commissioned officers the file closers.

The officer commanding verifies the alignment of the file closers; the officers and file closers cast their eyes to the front as soon as their alignment is verified.

At the command *front*, the men cast their eyes to the front and drop the left hand.

179. The officer commanding takes post facing to the front, three paces in front of the right guide, and as the inspector approaches, he faces to the left, commands:

1. *Inspection*, 2. POUCHES,

and facing to the front salutes him.

At the second command the pouches are shifted under the right arm to the front, the flap opened and held by the left hand, fingers extended, palm against the body, so that the flap strap covers the line of buttons, right hand at side.

As soon as inspected, the officer commanding returns sword and accompanies the inspector. When the latter begins to inspect the rank, the junior officers face about and stand at ease, sword at the order.

Commencing on the right, the inspector now proceeds to minutely inspect the pouch, accouterments, and dress of each soldier in succession.

After the inspector has passed each man closes and replaces the pouch.

180. The inspection being completed, the junior officers come to attention, carry sword and face to the front; the officer commanding again takes his post on the right, draws his sword, and facing to the left commands:

1. *Close ranks*, 2. MARCH.

At the command *march*, the junior officers face about and resume their posts in line of file closers; the file closers close to two paces from the rank. The officer commanding may direct the junior officers to stand fast in front of the detachment.

181. If the detachment, or part of it, should be mounted, or armed with revolvers, it will be inspected in this respect in accordance with cavalry drill regulations.

The clothing roll will be inspected, unrolled, on the soldier's bunk at inspection of quarters.

INSPECTION OF LITTERS.

182. The detachment being in line with strapped litters at the carry, the commands are given:

1. *Litters left*, 2. MARCH, 3. HALT.

1. *Inspection*, 2. LITTERS.

At *litters*, No. 1 of each squad steps back in line with No. 3, the litters are opened, held suspended until inspected, and then lowered, when the squads take posts at litters.

INSPECTION OF AMBULANCES.

183. The ambulances being in line at intervals of ten paces, with seats packed, each with a squad at ambulance posts, the commands are given:

1. *Inspection*, 2. AMBULANCES;

when each squad steps back three paces in rear of its ambulance. The inspector first examines the animals and harness, then the ambulance and contents, after which he directs the seats to be prepared, or such other work to be done as he desires executed.

MUSTER.

184. All stated musters of the detachment are, when practicable, preceded by a minute and careful inspection.

The detachment being in line with ranks open, the officer commanding, upon intimation of the mustering officer, commands:

Attention to muster.

He then returns sword, and hands a roll of the Hospital Corps detachment, with a list of absentees, to the mustering officer. The latter calls over the names on the roll; each man, as his name is called, answers, "Here," and steps forward one pace. The muster completed, the ranks are closed and the detachment dismissed.

After mustering, the presence of the men reported in the hospital or on duty is verified by the mustering officer, who is accompanied by the officer commanding.

TENT DRILL AND PACKING.

185. The canvas of a field hospital consists of hospital tents, conical wall tents, and common tents. The hospital tents are intended for use as wards and dispensary, the conical wall tents as squad and mess tents, and the common tents as latrine covers.

Tentage for medical officers is not included in that for the field hospital. Each medical officer is allowed one wall tent complete.

HOSPITAL TENT.

186. A hospital tent is 14 feet long, 15 feet wide, and 11 feet to ridge, the wall being $4\frac{1}{2}$ feet high; it furnishes comfortable accommodations for six patients, and requires to pitch it a ridge-pole and two upright poles, seven long tent pins on each side for the guy ropes and two on each side for the long guys, eighteen in all. Twenty-four small pins are needed for the front, rear, and walls.

187. The hospital tents should always be pitched first in the field hospital. One squad (4 men) under the direction of a non-commissioned officer is required. Nos. 1 and 3 work in rear, Nos. 2 and 4 in front (positions as *by four, prepare to lift*, par. 77).

188. At the commands:

1. *Pitch hospital tent*, 2. MARCH,

Nos. 1 and 2 procure canvas, Nos. 3 and 4 the poles. Nos. 3 and 4 place the ridgepole on the ground as directed, and the uprights in position, usually on the side opposite that from which the wind blows; then get each two mauls, nine large and twelve small pins, which they drop at their respective ends of the tent; after which they set a small pin at each end of the ridge to mark the rear and front openings. Meanwhile, Nos. 1 and 2 unroll the tent and spread it out on both sides of the ridgepole.

Nos. 1 and 3 in rear, Nos. 2 and 4 in front, slip the pins of the uprights through the ridgepole and tent. The fly (if used) is now placed in position over tent and the loops of the long guys over the front and rear pole pins. No. 1 secures center (door) loops over center pin in rear, and No. 4 in front, and each goes to his corner, No. 1 right rear, No. 2 right front, No. 3 left rear, No. 4 left front. All draw bottom of tent taut and square, the front and rear at right angles to the ridge, and fasten it with pins through the corner loops; then stepping outward two paces from the corner pins and one pace to the front (Nos. 2 and 4) or rear (Nos. 1 and 3), each securely sets a long pin, over which is passed the extended corner guy rope. Nos. 1 and 3 now go to rear, Nos. 2 and 4 to front pole and raise the tent to a convenient height from the ground, when Nos. 2 and 3 enter and seize their respective poles, and all together raise the tent until the upright poles are vertical. While Nos. 2 and 3 support the poles, Nos. 1 and 4 tighten the corner guys, beginning on the windward side. The tent being thus temporarily secured, all set the guy pins and fasten the guy ropes, Nos. 1 and 2 right, Nos. 3 and 4 left, and then the wall pins.

189. A wall tent or common tent is pitched in the same manner as a hospital tent. Care must be taken that the tent is properly squared and pinned to the ground at the door and four corners before being raised.

CONICAL WALL TENT.

190. The conical wall tent is $16\frac{5}{12}$ feet in diameter and 11 feet to the peak. It is provided with a hood, and will comfortably accommodate ten men, and may be made to hold twice that number.

To pitch it requires a tripod, pole, forty-eight tent pins, and a squad (four men) under the direction of a non-commissioned officer. No. 1 works on the right, No. 4 on the left, No. 2 in front, No. 3 in rear (positions as *at litter*, par. 68).

191. At the commands:

1. *Pitch conical wall tent*, 2. MARCH,

Nos. 1 and 2 procure canvas, No. 3 tripod and pole, and No. 4 the tent pins and two mauls. They all unroll the tent and spread it out near where it is to be pitched, apex at its center. No. 1 having taken a maul and three pins steps off eight paces directly outward (right or left) from the front corner of the hospital tent, on a line with its front, and drives two pins 2 feet apart, to mark the door of the conical wall tent; he then measures with the tent pole from the middle point between these pins directly backward, the far end of the pole determining the center of the tent, which he also marks with a pin; No. 3 places the tripod opened out flat, with ring over the center pin, and lays the pole on the ground, pin end at center pin. All now being at their posts, bring the canvas over the tripod till its center comes to the center pin and door at the front pins, when No. 2 slips the wall loop at each side of door over front pins. This is an important duty and upon its proper performance depends the proper pitching of the tent.

Nos. 1 and 4 commencing at rear and front of tent respectively, and working to the right and left, scatter the pins and pull out the guy ropes. Nos. 2 and 3 take each a maul, and commencing front and rear respectively, work right and left of the tent, driving the guy pins, placing them about one yard from the edge of the tent, each on a line with a seam. As the pins are driven, Nos. 1 and 4 place the ends of the guy ropes over them, working on their respective sides. When the pins are set, No. 2 creeps under the canvas, slightly raises the tent and places the pin of the pole through the plate attached to the chains at the top of the tent, and raising the pole, sets it in the ring of the tripod; No. 3 having, from the outside, placed the hood over the pole pin, enters the tent by creeping under, and assists No. 2 in raising the tripod, which being done, Nos. 1 and 4 tighten the guys; they then scatter the wall pins. The tent having been secured, Nos. 2 and 3 now take their posts outside and drive the wall pins, working as before, No. 2 toward the right rear, and No. 3 toward the left front; Nos. 1 and 4 straighten the tent and fasten the hood guys.

In pitching, as soon as any man has completed his assigned work, he assists the others until all have finished.

192. The tents having been pitched, they should be thoroughly ditched as soon as convenient.

193. To strike a tent: At the commands *strike tent, march*, the men take their posts; they first remove the wall pins, and then all the guy pins on their respective sides, except the four corner pins of the square tent, or the quadrant pins of the conical wall tents. Standing at their respective posts they remove the corner, or quadrant, guys from the pins and hold the tent until the command *Down* is given, when the tent is lowered to the indicated side. The canvas is then rolled up and tied by Nos. 1 and 2, while Nos. 3 and 4 fasten the poles, or tripod and pole, together, and collect the pins.

PACKING.

194. The pack equipment complete consists of—
1 combination halter and bridle, with leading line;
1 breast strap and chains;
1 breaching strap and chains;
1 pack saddle, with parts as follows, viz.:
    1 iron yoke frame;
    2 wooden side pieces;
    2 side pads of leather, lined with blanket and stuffed with hair;
    4 small straps to fasten saddle and pads together;
    2 pairs of quarter straps (each connected by a cross strap) with rings and two cincha (latigo) straps for each side;
    2 webbing cinchas;
    1 surcingle (cargo cincha);
    2 pairs of sling ropes;
    2 saddle blankets;
2 canvas chest covers.

The purpose of this equipment is to permit of the packing of the medical (No. 1) and surgical (No. 2) chests, in the event of wheeled transportation being impracticable.

It will be observed that the chests are so arranged upon the saddle as to permit of immediate access to their contents, to facilitate which the animal carrying the pack must habitually be led.

One litter squad will be designated as packers, who will see that the equipment is accurately fitted to the animal, and will be responsible for its care and condition.

No. 1 brings the saddle and puts it on from the near side; No. 2 bridles and blindfolds the animal, and holds him, taking care that he is never moved without first removing the blinder. He will also assist No. 1 in saddling, working on the off side.

TO SADDLE.

195. Place the folded saddle blankets one above the other, so that their front edges will come $2\frac{1}{2}$ inches in front of where the pommel end of the saddle is to rest; take the saddle by both yokes and place it squarely in position, a little in rear of its proper place; place the crupper under the dock and gently move the saddle forward into position, taking care not to disarrange or move the blankets; pass the latigo strap through the free end of the front cincha and tighten and secure it; then secure the rear cincha in the same manner, taking care that the rings of the cinchas, when cinched, are above the lower edge of the pads. Place the breast strap in position, and fasten the chains of the breast and breeching straps to the saddle. Pass the strap end of the cargo cincha, to its mid-length, under the yoke bars of the saddle, and throw both ends to the rear, off of the saddle, taking care that the bight of the cincha remains between the hooks of the saddle yoke.

TO LOAD.

196. Nos. 3 and 4 bring the chests, and, beginning with the medical (No. 1) on the near side, hang them upon the saddle, suspended from the yoke hooks by two rings permanently fastened to the back of each chest.

No. 1 having steadied the saddle by supporting the medical (No. 1) chest until the surgical (No. 2) chest is in place, then secures the load as follows:

No. 4 from the off side takes the strap end of the cargo cincha, brings it over No. 2 chest, and passes it under the animal's belly to No. 1, who, in the meantime, has brought the buckle end of the cincha over No. 1 chest; No. 1 then passes the strap through the buckle, and, with the assistance of No. 4, draws the cincha snug and buckles it.

TRANSPORTATION OF THE WOUNDED.

Anything other than medical or surgical chests are packed by means of the sling ropes.

No. 1 passes the sling ropes over the saddle to the off side, until the cross strap, placed mid-length of the ropes and joining them, comes parallel to the animal's spine; Nos. 3 and 4 place the off-side pack well up on the saddle, where No. 3 supports it with the left shoulder and throws the ends of the sling ropes over his right shoulder, in readiness to pass them over the pack; No. 4 then passes to the near side and assists No. 1 to place near-side pack well up on saddle, its edge, if possible, overlapping the upper edge of the off-side pack, where he supports it; No. 1 takes the end of the front rope and slipping it through loop passed to him by No. 3, secures it, then passes to the off side, secures the rear rope, the loop of which is passed to him by No. 4. The packs having been slung, are balanced, when No. 1 secures them with the cargo cincha, passed over the pack and under the animal's belly.

197. Scheme for packing hospital corps pouch.

Rear (in loops).			
Case with scissors, pins, etc.	Ball of wire gauze.	Flask with ammoniæ spiritus aromaticus.	Rubber tourniquet. Knife.

Front.

Packet.	Packet.	Packet.
Packet.	Packet.	Packet.

Bottom.

Six roller bandages.	Spool plaster.

TRANSPORTATION OF THE WOUNDED. 291

198. SCHEME FOR PACKING ORDERLY POUCH.

REAR (chiefly in loops).							
Chloroform in case.	Roll wire gauze.	Spools adhesive plaster.			Hypodermic syringe.	Mist. chloroformi et opii, in case.
^	^	Bottle catgut ligatures.	Ammoniæ spiritus aromaticus, in flask.	^	^
FRONT.							

Scissors.	Packet. Packet. Catheter, Pins. Diagnosis in case. tags. Packet. Pocket Packet. case.	Knife.

IN TRAY AT BOTTOM.

Four packages gauze.   Rubber tourniquet.
Six roller bandages.   Antiseptic tablets.

CLOTHING ROLL.

199. The articles heretofore carried in the knapsack or blanket bag will, together with the overcoat, be rolled in the piece of shelter tent supplied each soldier, and carried in the transportation wagon, or in the ambulance when no other transportation is provided. When the soldier is mounted and no wheel transportation is available, they will be carried on the saddle, as directed in the drill regulations for the cavalry.

Contents:
 1 woolen blanket;
 1 blue flannel shirt;
 1 undershirt;
 1 pair drawers;
 2 pair socks;
 1 towel;
 5 shelter-tent pegs;
 2 shelter-tent poles;
 1 overcoat.

The roll, which should when completed be 26 inches in length, is packed as follows:

First, spread out the blanket upon the ground and turn in its sides, making them overlap in such manner that the blanket, when so folded, shall equal in width the length of the longer shelter-tent pole. Double the blanket lengthwise, bringing the upper end 18 inches short of the lower end. Upon the doubled end of the blanket place, in the following order, the flannel shirt, undershirt, socks, drawers, and towel, so folded as to equal width of blanket.

Next, arrange pegs of shelter tent at upper end of clothing, three on one side and two on the other, points inward, bases flush with outer edge of clothing. On these place the shoes, one on each side, soles up, toes inward. Now roll tightly (beginning at bundle of clothing) as far as the blanket is doubled. Turn up the remaining 18 inches of the blanket and pull the upper thickness of this end over the bundle, thus securing it. The poles are now laid upon the bundle, and the overcoat, folded with its inside outward in such manner as to equal the width of the previous bundle, is rolled round the latter. Finally, the roll is completed by spreading out the shelter-tent half, folding in the rear flap, placing the bundle upon the flap, turning in the sides of the tent and rolling tightly. The whole is now secured by the straps furnished for the purpose.

200. HOSPITAL CORPS BUGLE CALL:

POSITION OF THE MEDICAL OFFICERS, HOSPITAL CORPS DETACHMENT, AND AMBULANCES ON THE MARCH.

201. The position of the medical department of a marching command is immediately in rear of the rear company of the organization to which it pertains, and in front of the rear guard.

With each ambulance is a driver and an ambulance orderly.

In camp the ambulances and medical department wagons are parked near the field hospital, and not with the wagon train.

INDEX.

Acetic acid, poisoning by, 197.
Acids, poisoning by, 197.
Aconite, poisoning by, 189.
"Adam's apple," 39.
Air cells, or vesicles, 41.
Air, quantity of, respired, 43.
Alcohol, poisoning by, 190.
Alimentation, 45.
Alkalies, poisoning by, 198.
Ammonia, poisoning by, 198.
Antiseptics, 91.
Aponeurosis, 27.
Apoplexy, 170.
  treatment of, 172.
  heat, 175.
Arm-bone or humerus, 14.
Arm slings, 81.
Arsenic, poisoning by, 193.
Arteries, anatomy of, 36.
  nutrient, 3.
Artificial respiration, 181.
  Hall's method, 185.
  Howard's method, 183.
  Sylvester's method, 181.
Asphyxia, 178.
  precautions in rescuing, 178.
  treatment of, 178.
Atropine, poisoning by, 190.
Axillary artery, compression of, 129.

Bandages, 68.
  application of, 69, 71.

Bandages, circular, 70.
  cravat, 83.
  double-headed knotted, 73.
  Esmarch or triangular, 76.
  Esmarch, a triangular, for chest, 79.
  for foot, 83.
  for hand, 78.
  for head, 78.
  for hip, 82.
  for shoulder, 78.
  figure-of-8, 71.
  four-tailed, 74.
  head, 73.
  hip spica, 72.
  large, square handkerchief for head, 76.
  materials for, 68.
  method of rolling, 69.
  roller, 68.
  shoulder spica, 73.
  six-tailed, 75.
  sling, 81.
  spiral reverse, 70.
  triangular, 76.
Basin, the, or pelvis, 18.
Bed-sores, 118.
Belladonna, poisoning by, 190.
Bladder, urinary, 55.
Blood, 30.
  amount of, in human body, 31.
  arterial, 35.
  circulation of, 34.

Blood, coagulation of, 31.
　composition, 30.
　corpuscles, 30.
　venous, 34.
　vessels, anatomy of, 36.
Bone, cancellous tissue, 2.
　compact tissue, 2.
　composition of, 1.
　hyoid, 10.
　innominate, 19.
　marrow of, 3.
　necrosis of, 3.
　oil of, 3.
　spongy tissue, 2.
Bones, classification of, 4.
Bony landmarks, artificial, 7.
Brachial artery, compression of, 130.
Brain, 59.
　compression of, 170.
　concussion of, 169.
　weight of, 63.
Brandy, 93.
Bread poultices, 89.
Breast-bone, or sternum, 11.
Bromine, solution of, 105.
Bronchial tubes, 41.
Bruises, 107.
Burns, 156.
　constitutional treatment of, 159.
　of first degree, 156.
　　treatment of, 157.
　of second degree, 157.
　　treatment of, 158.
　of third degree, 158.
　　treatment of, 158.

Camphor, poisoning by, 190.
Cantharides, poisoning by, 194.
Capillaries, anatomy of, 37.
Carbolic acid, or phenol, 93.
　poisoning by, 197.
Carpus, or wrist, 18.
Cartilage, or gristle, 24.

Catching fire, 159.
Caustic potash, poisoning by, 198.
Caustic soda. poisoning by, 198.
Cerebellum, 62.
Cerebro-spinal system, 59.
Cerebrum, 61.
Chafing, 205.
Chest, or thorax, 10.
　bandages, 79.
Chloral, poisoning by, 191.
Chloroform, poisoning by, 191.
Clavicle, or collar-bone, 12.
Colic, kidney, 55.
Collapse, 162.
Collar-bone, or clavicle, 12.
Common carotid artery, compression of, 127.
Compresses, 86.
Compression of axillary artery, 129.
　of brachial artery, 130.
　of brain, 170.
　　treatment of, 170.
　of common carotid artery, 127.
　of femoral artery, 132.
　of popliteal artery, 132.
　of radial and ulnar arteries, 131.
　of subclavian artery, 128.
Concussion of the brain, 169.
　treatment of, 169.
Contusions, 107.
Convulsions of children, 199.
Copper, poisoning by, 194.
Corn-meal poultices, 89.
Corrosive sublimate, or bichloride of mercury, 92.
　poisoning by, 197.
Cranial bones, fracture of, 142.
Cravat bandages, 83.
Creasote, poisoning by, 197.
Croton-oil, poisoning by, 194.
Cuticle, 56.

Deodorants, 105.
Derma, or true skin, 57.

INDEX.

Diaphragm, 42.
Digestion, 45.
Digitalis, poisoning by, 191.
Diploë, 2.
Disinfectants, 93.
  bichloride solution, 96.
  carbolic-acid solution, 96.
  dry chloride of lime, 97.
  heat, 96.
  milk of lime, 96.
  soup-suds solution, 95.
  strong soda solution, 98.
Disinfection, methods of effecting, 100.
  of closets and sinks, 99.
  of clothing, towels, etc., 97.
  of dead body, 100.
  of discharges, 98.
  of dishes, knives, etc., 99.
  of food and drink, 98.
  of hands and person, 97.
  of rags, cloths, etc., 100.
  of rooms and contents, 100.
  of sputum from consumptives, 99.
Dislocations, 153.
  of humerus, 153.
  of lower jaw, 154.
  of phalanges, 154.
Dog-bite, treatment of, 115.
Drill regulations for the hospital corps, U. S. Army, 226.
  Alignments, 228.
  Ambulance, the, 276.
  Detachment, the, 226.
  Inspection, 281.
  Litter drill, 241.
    marchings with, 251.
    the loaded, 254.
    improvisation of, 265.
  Marchings, 228.
    with litter, 251.
  Muster, 281.
  Rests, the, 240.

Drill regulations for the hospital corps, U. S. Army.
  Wounded, methods of removing, without litters, 268.
  Tent drill, 284.
    Clothing roll, 292.
    Load, to, 288.
    Packing, 287.
    Saddle, to, 288.
    Tent, conical wall, 285.
      hospital, 284.
Drowning, 179.
  treatment of, 180.
  Hall's method, 184.
  Howard's method, 183.
  Sylvester's method, 181.

Ear, foreign bodies in, 201.
Emetics, 187.
Endocardium, 36.
Endosteum, 4.
Epidermis, 56.
Epiglottis, 39.
Epilepsy, 173.
  treatment of, 173.
Epistaxis, or nose-bleed, 126.
Eucalyptus, oil of, 93.
Excretion, 29.
Eye, foreign bodies in, 200.

Fainting, 166.
  treatment of, 167.
Falling-sickness, 173.
Femoral artery, compression of, 132.
Femur, the, or thigh-bone, 19.
  fractures of, 149.
Fever, sun, 175.
Fibula, the, or splint-bone, 22.
  fractures of, 151.
Fingers, dislocation of, 154.
  fractures of, 148.
Fits, epileptic, 173.
Flaxseed poultices, 88.
Fontanelles, the, 9.

Food, ration of, 216.
Foot bandages, 83.
Foot soreness, 205.
Forearm, fractures of, 146.
Foreign bodies in ear, 201.
  in eye, 200.
  in larynx, 203.
  in nose, 202.
  in pharynx, 203.
Fractures, 136.
  classification of, 136.
  diagnosis of, 137.
  treatment of, 138.
  of clavicle, 145.
    treatment of, 145.
  of cranial bones, 142.
  of femur, 148.
    treatment of, 149.
  of forearm, 146.
    treatment of, 147.
  of forearm, middle of, 147.
    treatment of, 147.
  of humerus, 146.
    treatment of, 146.
  of inferior maxillary bone, 142.
  of leg, 151.
    treatment of, 152.
  of metacarpal bones, 147.
    treatment of, 148.
  of metatarsal bones, 153.
  of patella, 152.
    treatment of, 153.
  of phalanges of fingers, 148.
    of toes, 153.
  of ribs, 143.
    treatment of, 144.
  of scapula, 146.
    treatment of, 146.
  of spinal column, 143.
    treatment of, 143.
  of tibia and fibula, 151.
  of vertebral column, 143.
  splints for, 140.
Frost-bite, 159.

Frost-bite, treatment of, 160.
"Funny-bone," 18.

Gangrene, 117.
  treatment of, 118.
Gastric juice, function of, 49.
Gland, definition of, 29.
Glands, salivary, 47.
  sweat, 57.
Gravel, 54.
Gristle, or cartilage, 24.
Gullet, or œsophagus, 47.

Hæmatemesis, or hæmorrhage from stomach, 135.
Hæmoptysis, or "spitting of blood," 134.
Hæmorrhage, arrest of, 120.
  classification of, 120.
  treatment of, after extraction of teeth, 126.
  arterial, 121.
  capillary, 125.
  nasal, 126.
  venous, 124.
  from axillary artery, 129.
  from brachial artery, 130.
  from common carotid artery, 127.
  from femoral artery, 132.
  from lips, 126.
  from mouth, 126.
  from popliteal artery, 132.
  from radial artery, 131.
  from scalp, 126.
  from subclavian artery, 128.
  from ulnar artery, 131.
  secondary, 132.
    symptoms of, 133.
    treatment of, 134.
Hair-follicles, 57.
Halstead's litter, 222.
Hand bandages, 78.
"Haunch," or innominate bone, 19.

Head bandages, 78.
Heart, 32.
  anatomy of, 33.
  power of, 36.
Heat, dry, application of, 90.
  moist, application of, 90.
Heat-stroke, 175.
Hip bandages, 82.
Hominy poultices, 89.
Humerus, or arm-bone, 14.
Hydrocyanic acid, poisoning by, 192.
Hygiene, 207.
  air, 218.
    effects of bad, 219.
  baths, 207.
    addition to, of alcohol, etc., 208.
    Russian or Turkish, 208.
    sea, 208.
    warm, 208.
  clothing, cotton and linen, 210.
    woolen, 209.
  dandruff, 209.
  exercise, 221.
  food, 211.
    constituents of, 212-215.
  scurvy, 215.
  skin, activity of, 207.
  soap, the use of, 209.
  water, 216.
    amount required daily, 216.
    lake, 217.
    rain, 216.
    spring, 216.
    well, 217.
Hyoid bone, 10.
Hysteria, 174.
  treatment of, 174.
Hysterics, 174.

Inferior maxillary bone, dislocation of, 154.
  fracture of, 142.

Innominate, or "haunch" bone, 10.
Insolation, 175.
Instep, or tarsus, 22.
Integument, or skin, 56.
  appendages of, 57.
  care of, 57.
  functions of, 57.
Intestine, large, 50.
Intestine, small, 49.
Intoxication, 172.
  treatment of, 172.
Iodine, poisoning by, 195.
Iodoform, 93.
Irritant poisons, 193.

Joints, classification of, 23.
  composition of, 24.
  movements of, 23.

Kidney colic, 55.
Kidneys, 54.
Knee-cap, or patella, 22.
Knots, 85.

Landmarks, superficial bony, 7.
Larynx, 39.
  foreign bodies in, 203.
Laudanum, poisoning by, 192.
Ligaments, 24.
Lips, hæmorrhage of, 126.
Litter, Halstead's, 222.
  extemporized, 225.
  manufactured, 222.
Liver, 51.
  functions of, 52.
Lower jaw, dislocation of, 154.
  fracture of, 142.
Lungs, 42.
  capacity of, 44.
Lye, poisoning by, 198.
Lymphatics, 4.

Marrow of bone, 3.
Marsh's stretcher, 225.

Mastication, 45.
Matches, poisoning by, 196.
Membranes, mucous, 29.
    serous, 29.
    synovial, 25.
Mercury, bichloride of, 92.
Metacarpal bones, 18.
    fracture of, 147.
Metacarpus, the, 18.
Metatarsal bones, 22.
    fracture of, 153.
Milk, sterilization of, for infants, 102.
Morphine, poisoning by, 192.
Mouth, hæmorrhage of, 126.
Mucous membranes, 29.
Muriatic acid, poisoning by, 197.
Muscles, 25.
    atrophy of, 28.
    involuntary, 28.
    rigidity of, at death, 28.
    voluntary, 25.
Mushrooms, poisoning by, 192.
Mussels, poisoning by, 192.
Mustard poultices, 89.

Narcotic poisons, 189.
Necrosis of bone, 3.
Nerves, 63.
    cranial, 64.
    motor, 64.
    sensory, 64.
    spinal, 66.
    sympathetic, 66.
Nervous system, 59.
Nitric acid, poisoning by, 197.
Nose-bleed, or epistaxis, 126.
Nose, foreign bodies in, 202.
Nux vomica, poisoning by, 195.

Œsophagus, or gullet, 47.
Opium, poisoning by, 192.
Organ, definition of, 25.
Oxalic acid, poisoning by, 197.

Padding for splints, 141.
Pancreas, 52.
    function of, 53.
Paralysis, stroke of, 170.
Paris green, poisoning by, 193.
Patella, the, or knee-cap, 22.
Pearlash, poisoning by, 198.
Pelvis, the, or basin, 18.
Pericardium, 34.
Periosteum, 3.
Phalanges, 18, 22.
    dislocation of, 154.
    fractures of, 148.
Pharynx, or throat, 47.
    foreign bodies in, 203.
Phenol, 93.
Phosphorus, poisoning by, 196.
Pleura, 42.
Plugs or tampons, 87.
Poisoning, 186.
    treatment of constitutional, 189.
    local, by emetics, 189.
    by stomach-pump, 188.
Poisoning by acetic acid, 197.
    by aconite, 189.
    by alcohol, 190.
    by ammonia, 198.
    by arsenic, 193.
    by atropine, 190.
    by belladonna, 190.
    by camphor, 190.
    by cantharides, 194.
    by carbolic acid, 197.
    by caustic potash, 198.
    by caustic soda, 198.
    by chloral, 191.
    by chloroform, 191.
    by copper, 194.
    by corrosive sublimate, 197.
    by creasote, 197.
    by croton-oil, 194.
    by digitalis, 191.
    by hydrocyanic acid, 192.
    by iodine, 195.

Poisoning by laudanum, 192.
  by lye, 198.
  by matches, 196.
  by morphine, 192.
  by muriatic acid, 197.
  by mushrooms, 192.
  by mussels, 192.
  by nitric acid, 197.
  by nux-vomica, 195.
  by opium, 192.
  by oxalic acid, 197.
  by Paris green, 193.
  by pearlash, 198.
  by phosphorus, 196.
  by poison oak or ivy, 198.
  by prussic acid, 192.
  by salts of lemon and sorrel, 197.
  by Spanish fly, 194.
  by strychnine, 195.
  by sulphuric acid, 197.
  by tartar emetic, 196.
  by zinc, 196.
  treatment of, 186.
Poison oak or ivy, poisoning by, 198.
Poisons, classification of, 186.
  irritant, 186, 193.
  narcotic, 186, 189.
Potash, caustic, poisoning by, 198.
Poultices, 87.
  bread, 89.
  corn-meal, 89.
  flaxseed, 88.
  hominy, 89.
  mustard, 89.
Prussic acid, poisoning by, 192.

Radial artery, compression of, 131.
Radius, 14.
Respiration, 39.
Ribs, 11.
  fracture of, 143.

Salivary glands, 47.
Salts of lemon, poisoning by, 197.
Scalds, 159.
Scalp, hæmorrhage of, 126.
Scapula, or shoulder-blades, 12.
Secondary hæmorrhage, 132.
  constitutional symptoms of, 133.
  treatment of, 134.
Secretion, 29.
Serous membranes, 29.
Shin-bone, or tibia, 21.
Shock, 162.
  treatment of, 163.
Shoulder bandages, 78.
Shoulder-blade, or scapula, 12.
Skeleton, the, 4.
Skin, or integument, 56.
  appendages of, 57.
  care of, 57.
  false, 56.
  functions of, 57.
  true, 56.
Skull, division of bones of, 8.
Sling bandages, 81.
Snake-bite, treatment of, 115.
Sorrel, poisoning by, 197.
Spanish fly, poisoning by, 194.
Spinal cord, 62.
Spine, the, 6.
  changes in, 6.
"Spitting of blood," or hæmoptysis, 134.
Spleen, 58.
Splint-bone, the, or fibula, 22.
Splints, 140.
  padding for, 141.
Sprains, 155.
  treatment of, 155.
Sterilization of milk for infants, 102.
Sternum, the, 11.
Stomach, 48.
  anatomy of, 48.
  capacity of, 48.

Stretcher, essentials of, 222.
  Halstead's, 222.
  Marsh's, 225.
Stroke of paralysis, 170.
Strychnine, poisoning by, 195.
Stunning, 169.
Suffocation, 178.
Sulphuric acid, poisoning by, 197.
Sunstroke, 175.
  treatment of, 176.
Sweat-glands, 57.
Sweet-breads, 52.
Sympathetic system, 66.
Syncope, 166.
  treatment of, 167.
Synovial membrane, 25.

Tampons, or plugs, 87.
Tarsus, or instep, 22.
Tartar emetic, poisoning by, 196.
Teeth, 45.
  hæmorrhage following extraction of, 126.
Tendons, 27.
Tetanus, or lockjaw, 199.
Thigh-bone, or femur, 19.
Thorax, or chest, 10.
Throat, or pharynx, 47.
  foreign bodies in, 203.
Thymol, 93.
Tibia, or shin-bone, 21.
Toes (phalanges), fracture of, 153.
Trachea, or windpipe, 39.
Transportation of wounded, 222.
  (See also Drill regulations for the hospital corps, U. S. Army.)

Ulna, 16.
Ulnar artery, compression of, 131.
Unconsciousness, 161.
Ureters, 54.
Urine, 54.
  quantity secreted, 57.

Veins, anatomy of, 38.
Vertebræ, the, 6.
Vertebral column, curves of, 7.
  division of, 7.
  fracture of, 143.
Vesicles, or air-cells, 41.
Vitriol, oil of, poisoning by, 194.

Whisky, 93.
Windpipe, or trachea, 39.
Wounded, transportation of, 222.
  (See also Drill regulations for the hospital corps, U. S. Army.)
Wounds caused by insects, 116.
  classification of, 109.
  contused, 111.
  gangrenous, 117.
  gunshot, 109.
  healing of, 111.
  incised, 109.
  lacerated, 109.
  poisoned, 110.
  punctured, 109.
  of abdominal walls, 116.
  of thorax, 117.
  treatment of, 112.
Wrist, or carpus, 18.

Zinc, poisoning by, 196.

THE END.

PRACTICAL DIETETICS,

With Special Reference to Diet in Disease.

By W. GILMAN THOMPSON, M. D.,

Professor of Materia Medica, Therapeutics, and Clinical Medicine in the University of the City of New York; Visiting Physician to the Presbyterian and Bellevue Hospitals, New York.

Large 8vo.   Eight Hundred Pages.   Illustrated.

Cloth, $5.50; sheep, $6.00.

SOLD ONLY BY SUBSCRIPTION.

"We commend to the critical attention of the medical profession this new and valuable work. We hesitate not to express the conviction that within the pages of this volume will be found more of explicit, reliable, and practical instruction with reference to the selection, preparation, and administration of foods appropriate to every age and condition, both in sickness and in health, than has hitherto been presented in any form to the medical profession."—*North American Practitioner.*

"The arrangement of this treatise is such that a busy practitioner can turn in a moment to a dietary which is adapted to any disease he may have under treatment and there find specific directions. We commend the book to every physician, believing that its frequent use will relieve him from a part of his professional work which has in the past been most unsatisfactory both to himself and to his patient."
—*Brooklyn Medical Journal.*

"This is at once the best and most exhaustive book upon this subject with which we are familiar. The best, because, in the first place, it is written by a teacher of therapeutics who knows the needs of the practicing physician, and yet who has taught in previous years as a professor of physiology all that one needs to know in regard to the principles of digestion and assimilation. For this reason the author is unusually well qualified to prepare a useful manual, but it is not until one has perused the volume that he thoroughly grasps the scope and depth of the manner in which Dr. Thompson has treated his subject."—*Therapeutic Gazette.*

"Without any exception we believe this work to be one of the best, if not the best, for practical usefulness that has issued from any press by any author in the last ten years. It is particularly useful because it supplies a vacancy in the library which every physician finds whenever he has a case to treat, and where diet occupies a part in the treatment and the recuperation of the patient. . . . It is complete in every department, each chapter being a model of conciseness and perfectness. With a book like this at hand, many a day's sickness will be prevented by the attending physician being able to prescribe a proper diet."—*Medical Current.*

D. APPLETON AND COMPANY, NEW YORK.

TRAUMATIC INJURIES OF THE BRAIN AND ITS MEMBRANES.

With a Special Study of Pistol-Shot Wounds of the Head in their Medico-Legal and Surgical Relations.

By CHARLES PHELPS, M. D.,

SURGEON TO BELLEVUE AND ST. VINCENT'S HOSPITALS.

8vo. 582 pages. With 49 Illustrations. Cloth, $5.00.

SOLD ONLY BY SUBSCRIPTION.

"This work is a concise and systematic treatise on that division of brain surgery arising from injuries of the brain through external violence, and has been based almost exclusively on the observation of five hundred consecutive cases of recent occurrence. Although the cases were so numerous, it seems they were incomplete only in the illustration of secondary pyogenic infection, which is naturally a tribute to the skill of the surgeons in charge of the cases, inasmuch as they were kept from infection. This clinical deficiency has been supplied by excerpts from Macewen's work on Inflammations of the Membranes of the Brain and Spinal Cord, with the permission of that gentleman. We have no hesitation in saying that it is the most complete work on this division of brain surgery which has yet appeared in America."—*Journal of the American Medical Association.*

"This book will prove of great service to both physician and surgeon; and to those interested in medical jurisprudence it will be of incalculable value. The author is not embarrassed by his great wealth of material; he studies it exhaustively, and arranges it clearly, concisely, and with great care and discrimination. The chapters on Pistol-Shot Injuries are particularly instructive, and the series of experiments on cadavers replete with interest. One of the strongest features of the book is the large number of photographic representations of cranial injury."—*National Medical Review.*

"We have here a new work highly creditable to American authorship and adding a material contribution to our present literature upon brain surgery. The first part of the work is devoted to the consideration of traumatic lesions of the cranium and its contents, embracing their pathology, symptomatology, diagnosis, prognosis, and treatment. Part II is an exceedingly interesting and original discussion of medico-legal and surgical relations and treatment of pistol-shot wounds of the head. Part III contains a condensed history of three hundred cases of intracranial traumatism verified by necropsy. A most interesting feature of the work is the introduction of a large number of full-page photographic illustrations of the effects of pistol-shot wounds produced by those of different calibers and at different distances. The work will at once be appreciated as one of original investigation, and especially by those who are particularly interested in brain surgery."—*North American Practitioner.*

D. APPLETON AND COMPANY, NEW YORK.

www.ingramcontent.com/pod-product-compliance
Lightning Source LLC
Chambersburg PA
CBHW030740230426
43667CB00007B/787